OUR SCIENCE, OURSELVES

A VOLUME IN THE SERIES
Activist Studies of Science & Technology

SERIES EDITOR
Sigrid Schmalzer

OUR SCIENCE, OURSELVES

HOW GENDER, RACE,

AND SOCIAL MOVEMENTS

SHAPED THE STUDY

OF SCIENCE

CHRISTA KULJIAN

University of Massachusetts Press
Amherst and Boston

Copyright © 2024 by University of Massachusetts Press
All rights reserved
Printed in the United States of America

ISBN 978-1-62534-818-0 (paper); 819-7 (hardcover)

Designed by Sally Nichols
Set in Adobe Jenson Pro
Printed and bound by Books International, Inc.

Cover design by adam b. bohannon
Cover art by noche + graficriver,
Science Laboratory and Women's Symbols. Shutterstock.com.

Library of Congress Cataloging-in-Publication Data

Names: Kuljian, Christa author.
Title: Our science, ourselves : how gender, race, and social movements
shaped the study of science / Christa Kuljian, University of
Massachusetts Press.
Description: Amherst : University of Massachusetts Press, [2024] | Series:
Activist studies of science & technology | Includes bibliographical
references and index. | Identifiers: LCCN 2024031464 (print) | LCCN 2024031465 (ebook) | ISBN
9781625348180 (paperback) | ISBN 9781625348197 (hardcover) | ISBN
9781685750893 (ebook) | ISBN 9781685750909 (epub)
Subjects: LCSH: Feminism. | Social movements. | Women—History.
Classification: LCC HQ1155 .K85 2024 (print) | LCC HQ1155 (ebook) | DDC
305.42—dc23/eng/20240802
LC record available at https://lccn.loc.gov/2024031464
LC ebook record available at https://lccn.loc.gov/2024031465

British Library Cataloguing-in-Publication Data
A catalog record for this book is available from the British Library.

*To all of the inspiring women
throughout time,
and in my life,
especially my sister Sarah*

CONTENTS

ILLUSTRATIONS

OUR SCIENCE, OURSELVES

INTRODUCTION

When I was growing up in greater Boston in the late 1970s, my parents bought me a copy of the revolutionary guide on women's health, *Our Bodies, Ourselves*, which gave me not only knowledge about my body, but also a sense of the burgeoning women's movement. The book was produced and published locally by the Boston Women's Health Book Collective in 1970, but it was reaching women throughout the country. I watched my mother, who had previously stayed at home taking care of me and my younger sister, graduate with a degree from Boston University and chart a new course as a rehabilitation counselor at my school. In the fall of 1980, when I was eighteen years old, I arrived on the Harvard College campus as a first-year student, and for a moment I mistakenly thought that, because the women's movement had peaked in the 1970s, it had largely subsided. I had a subscription to *Ms.* magazine, and I felt that my classmates and I had benefited from the women's movement and that it was now our responsibility to fulfill its potential. I soon learned, however, that there was a long way to go in terms of achieving justice for women and that social movements would continue to be a critical force in society.

Part of my education took place in a course taught by Professor Ruth Hubbard, who in 1974 had become the first woman to achieve tenure in biology at Harvard[1]. Hubbard had spent the 1940s through the 1960s as a research assistant in her husband's lab, conducting experiments to reveal the biochemical workings of the eye. After years of this, she was swept up in the second wave of the women's movement, which profoundly influenced her. For Hubbard, feminism and questioning the neutrality of science came together at the same time. She asked, If scientists were more representative of society in terms of gender, race, and class, would science look different? Moreover, why, after decades of important lab research, was she not tenured? And why did no one think this was a problem?

Born Ruth Hoffman, Hubbard had fled Vienna as a teenager with her family before World War II.[2] Growing up in a socialist home, she had always identified with left-leaning politics. Yet she admitted that it had taken two major social upheavals—the Vietnam War and the women's movement—to push her to step back from her lab work, reread Charles Darwin, and recognize that scientists were greatly affected by their social and personal background. In an interview years later, Hubbard said that the idea that science had been made almost exclusively by highly educated, economically privileged, white men hit her "like a bolt out of the blue." This revelation was one of her most liberating insights, and it caused her to start questioning the concept of scientific objectivity. She shifted her attention from looking closely at proteins in the retina to looking closely at scientists who were biased in their study of women's biology.[3]

Hubbard was in her late fifties when I took her then-famous course "Bio 109: Biology and Women's Issues." In corduroy pants and Birkenstocks, her salt-and-pepper hair pulled back into a long braid, she lectured on the social context of science and the social construction of women's biology, sexuality, sex and gender, sex differences, menstruation, women's health, and reproductive rights. She talked to us a great deal about scientists, including Darwin, who believed the Victorian stereotype of the active male and the passive female. Many of these scientists, she said, had imposed their beliefs onto their diverse research into animals, humans, algae, bacteria, and cells. She often described them as promoting "a self-fulfilling prophecy": looking for passive elements in nature and calling them female, looking for active dynamics in nature and calling them male. Hubbard helped me to see the tendency of male scientists to use biology to affirm patriarchy. If these revelations sound unremarkable in 2024, that is because of the work that she and many others did to develop feminist critiques of scientific knowledge. Back in 1983, however, they were revolutionary ideas. I had learned a new way to understand oppression.[4]

Influenced by Hubbard, my senior thesis for my degree in the history of science explored how the American medical profession presented women's biology between 1870 and 1920.[5] In the late 1800s, for instance, doctors were telling young white women to focus their energies on reproduction, not higher education. In *Sex in Education, or, A Fair Chance for the Girls* (1873), Edward H. Clarke, a respected physician and board member of the Harvard Medical School, argued that women should avoid academic study because it would negatively affect their reproductive system and their unborn children. As I wrote in a class essay, "what an ingenious way for the elite male community to keep women in 'their place.'" Clarke's argument focused solely on middle- and upper-class white women; it did not apply to working-class or enslaved women. In fact, he was a prime proponent of the concept of *race suicide*, which, in his view, required protecting and propagating white upper-class society. Thus, he believed, women from this class needed to focus on child bearing.

My research for my thesis focused on how certain women in medicine had pushed back against Clarke's theories. In Hubbard's course, I learned that it was important to explore the assumptions that inform a scientist's research questions and findings and to understand the social context that often shapes them. But Hubbard wasn't my official thesis supervisor. That was Margaret Rossiter, a visiting lecturer at Harvard who had just published her first book, *Women Scientists in America* (1982). Until this point, many people, even knowledgeable historians of science, had dismissed women's involvement in science. But Rossiter's book clearly showed that women had been active in the sciences in the United States for more than a century.

I was honored that she was watching over my work. In addition, one of Hubbard's teaching assistants, Dolita Cathcart, was a regular support as I researched and wrote my thesis. Cathcart was an African American pre-med student who went on to become a historian, and she pointed out to me that Rossiter's book had focused on white women, as did Bio 109. The course's reading list relied heavily on published anthologies that Hubbard had co-edited, including *Biological Woman: The Convenient Myth*, *Genes and Gender II*, and *Women's Nature*. Audre Lorde's *Cancer Journals* (1980) was one of a handful of works by women of color that appeared on the recommended-reading list. Another was an essay by Beverly Smith, one of the authors of the Combahee River Collective Statement. Both Lorde and Smith shared their experience of the medical establishment and Black women's health via an analysis of gender and race. Yet in the mid-1980s, thoughts on race and racism were marginal to the growing feminist critique of science.

The book in your hands, *Our Science, Ourselves*, grew from my formative experience in Hubbard's course, my thesis research into the history of science, and my conversations with Cathcart, but it also has roots in another era. In 2016, I published *Darwin's Hunch: Science, Race, and the Search for Human Origins* about the history of paleoanthropology in South Africa. The book explores the kinds of questions that Ruth Hubbard and my other history of science professors, Stephen Jay Gould and Everett Mendelsohn, often asked. How did the changing social and political context over the past century shape the search for human origins? What influence did colonialism and apartheid have on the search? And how did scientists' assumptions, especially about race, shape their research questions about human evolution?

While writing *Darwin's Hunch*, I was immersed in the impact of racism, sexism, and race typology in science in the twentieth century. Then, soon after the book appeared, I was struck by several pronouncements by scientists that were again bringing sexism and science to the fore. In July 2017, James Damore, a software engineer at Google, wrote in an internal memo that "the gender gap in tech" likely exists because of biological differences between men and women. Many scientists pushed back against Damore, and Google fired him for "advancing harmful gender stereotypes in our workplace." Not long afterward, the physicist Jess Wade argued that "those biases have calcified the idea that the inequalities that surround us are rooted in our biology rather than our society."[6]

Still, male scientists continued to promote such viewpoints. "Physics was invented and built by men," proclaimed Alessandro Strumia, a senior Italian scientist, during a September 2018 seminar on gender issues in physics at CERN, the European nuclear research center in Geneva. According to him, the low number of women in the field was proof that women were innately less capable than men; and he suggested that male scientists were being discriminated against in order to give opportunities to women.[7]

What was going on? These statements reminded me of what former Harvard president Larry Summers had said back in 2005. Speaking at a conference on women and people of color in science, technology, engineering, and mathematics, he'd declared that women were underrepresented in the sciences in large part because of biological and genetic sex differences between men and women—what he called a "different availability of aptitude."[8] Such pronouncements reminded me of Ruth Hubbard's course and how she had shown that society often shapes a scientist's assumptions. They also called to mind what Clarke had written in *Sex in Education* in 1873.

Why were these myths still having an impact today? I decided it was time to go back to my class notes and my thesis and look more closely at Hubbard's research. Who had she worked with at the time? What were other scientists with a feminist awareness saying in the 1970s and 1980s? Over the course of many interviews and much research into the archives, I discovered numerous feminist critiques of science that had developed during those decades and that continue to be relevant to science today.

In September 2016, Hubbard died at the age of ninety-two. That same month, Margot Lee Shetterly published *Hidden Figures*, later made into a film by the same name, which focused on African American women mathematicians, including Katherine Johnson, who played a critical role at NASA in the 1960s. Having recently published my own book about science and race, I knew it would be impossible to study gender and science without centering race as well. As I studied scholarship from the decades after I'd taken Hubbard's course, I learned that feminists of color were also making this point in the study of science. My research on gender and science was leading me into discoveries about race, class and sexuality.

When I began my research for this book in 2019, I learned about a feminist microbiologist named Dorothy Burnham, who had helped Ruth Hubbard understand that the scientific myths about Black women were different from those about white women. Burnham had always loved science. She had studied biology at Brooklyn College in the 1930s, and in the 1960s was the head of Staten Island Hospital's research laboratory. She also spent several years at Sloan Kettering, focusing on staphylococcal infections, contagion, and bacterial spread.[9]

Burnham was part of a small cohort of African American women in science before 1940, and her leadership in a lab was unusual for the time. She was also actively involved in civil rights activism in Birmingham, Alabama, and New York City. But in the 1940s and 1950s, women scientists such as Hubbard and Burnham isolated their politics from their science. Thus, despite her long commitment both to racial justice and science, Burnham did not bring these roles together until 1977, when she was sixty-two years old. Then, under the impetus of the women's movement and the formation of an organization called the Genes and Gender Collective, she spoke publicly about how the field of biology had contributed to scientific racism and sexism. Burnham recognized that the women scientists who were starting to speak out about scientific sexism were not considering scientific racism and how it was affecting Black women. From her experience, she knew it was critical to address both.[10]

In the 1960s and 1970s, Hubbard, Burnham, and other women scientists started focusing more and more on the ways in which biologists were promoting stereotypes of women. They felt compelled to speak about these issues, ask questions, write, and publish. What Hubbard's course represented, and what this book explores in detail, is the major impact that the second wave of the women's movement had on women scientists and philosophers of science in the United States. It also considers the feminist critiques of science that grew out of this engagement and led to the loose consolidation of a discipline known as feminist science studies.

The Boston area was an important hub for the feminist analysis of science, thanks to a major concentration of universities, colleges, students, and scientific research. Yet even though all of the seven women I highlight in this book worked in Boston, none of them grew up there. Ruth Hubbard was from Vienna, Austria. Rita Arditti grew up in Buenos Aires, Argentina. Evelyn Fox Keller, Nancy Hopkins, and Anne Fausto-Sterling were born in New York City. Evelynn Hammonds is from Atlanta, Georgia, and Banu Subramaniam is from Chennai, India. Each was drawn to the Boston area to study science, network with other scientists, or to take a research job at a local university.

Thus, in addition to focusing on individuals, this book considers what was happening around the Boston area in the 1970s and 1980s and how those events contributed to the development of feminist critiques of science. The paths and careers of these seven women overlapped or ran parallel in significant ways. During the 1970s, 1980s, and 1990s, they all knew one another and, with the exception of Nancy Hopkins, supported each other's writing, teaching, speaking, and critiquing. Some were also influenced by the women's groups and consciousness-raising groups that were popular at the time. Such gatherings gave them the impetus to begin thinking differently about their role in their field and to question the objectivity of science in new ways. They began to question and write about the ways in which they had been excluded from their field. They began to contemplate the challenges for women in science as well as issues linked to gender and science.

One of my major goals in this book is to chart the efforts that feminist science studies has made over the decades to take an intersectional approach, embracing not only gender but also race, class, and sexuality. The Combahee River Collective Statement in 1977 had an influence, as did Kimberlé Crenshaw's 1989 scholarship on intersectionality. A decade later, in 1999, Patricia Hill Collins wrote about the importance of intersectionality and scientific knowledge, a point that scholars today still struggle to make heard.[11]

Thus, as I discuss these scientists, I'll consider, too, the relationship between social movements and the development of feminist science studies: the role of the women's movement, the founding of Science for the People and the Genes and Gender Collective, the growth of Black feminism, and the impact of other groupings that propelled a feminist view of science. These organizations helped women look beyond the private and the personal and begin to examine broader social problems and to ask larger questions about science. They began to see that science was not isolated from politics but shaped by it.

Part 1 explores the history of women in science in Europe and the United States and introduces three of the book's main characters—Ruth Hubbard, Rita Arditti, and Evelyn Fox Keller—all of whom became scientists after World War II. Part 2 considers how women in groups such as Science for the People began to question scientific bias related to sex and gender. It introduces Evelynn Hammonds, and Anne Fausto-Sterling, who were inspired by Hubbard, Arditti, and Keller and the growing number of feminist scientists who were writing about science in new ways. It also introduces Nancy Hopkins, a scientist who, at the time, wanted to stay far away from feminists. Part 3 explores the impact of a growing feminist awareness on the understanding of science and introduces Banu Subramaniam, the youngest of the seven scientists, who arrived in the United States to study evolutionary biology in 1986.

The seven scientists featured in this book were crucial to the early history of feminist science studies. Their insights helped to create a set of tools and a field of study that has grown over the decades. Today, feminist science studies contributes to many fields, including cell biology, developmental biology, chemistry, physics, science and technology studies (STS), and the history of science. In many cases, the insights these women offered decades ago have become part of mainstream knowledge in their fields. And, like dendrites spreading impulses far beyond their source, they continue to inspire new research, knowledge, and writing that shapes our view of the world, science, and ourselves.

PART ONE
WOMEN IN SCIENCE

CHAPTER I

CENTURIES OF EXCLUSION

From the beginning, as western science was being developed in Europe, men excluded women from academies of learning, including Oxford, Cambridge, and other early universities. The Royal Society, founded in London in 1660 and one of the oldest scientific associations, did not elect a woman to full membership until 1945. The Academy of Sciences in France, established in 1666, did not accept a woman to full membership until 1979. In 1911, it even snubbed Marie Curie, the first person ever to win the Nobel Prize twice. The Academy of Science in Berlin had a similar trajectory.

Within these European academies, scientists came together to discuss the work that would become modern western science. Thus, by excluding women, science developed as if they were completely absent. Even when women, working outside the academies, made fundamental contributions, they were largely ignored.

As science developed, it also bolstered the demarcation of categories of difference, including sex, sexuality, race, and nationality. Notably, science, scientific sexism, and scientific racism all came into being in the seventeenth and eighteenth centuries, and scientists who were studying categories of difference often embraced prevailing social contexts and thinking that justified both patriarchy and white supremacy. Yet the choice to exclude women from the academies was not inevitable. Scholars who explore the roots of patriarchy in the sciences have shown that qualified women have long argued for a place in science, for a place in creating knowledge about the natural world.[1]

Like Europe, the young United States had numerous scientific societies, most of which excluded women. Among them were the American Academy of Arts and Sciences, established in 1780 in Cambridge, Massachusetts; the American Association for the Advancement of Science, formed in 1848 in Philadelphia; and the National Academy of Science, founded in 1863 in Washington, D.C. The organizers and members, who included luminaries such as John Hancock, offered countless reasons for excluding women and people of color from the sciences. In the eighteenth century, they described women as less intelligent than men because of their smaller brain size. In the nineteenth-century, they wrote that women could not pursue higher education because it would shrivel their wombs and their ovaries. In 1873, Edward H. Clarke, a well-known Boston physician, published his attack on higher education for women in *Sex in Education, or, A Fair Chance for the Girls*, describing women as physically delicate and unable to engage in science without significantly threatening their ability to bear children.

The astronomer Maria Mitchell, one of America's most prominent woman scientists, immediately refuted Clarke's claims. She had the stature to do so. In 1848, she had been the first woman member of the American Academy of Arts and Sciences, and she served as a professor at Vassar College for women. And she was not alone. In 1873, the same year that Clarke published his book, there were at least four hundred women science instructors in the United States, many of them working at fledgling women's colleges such as Smith, Mount Holyoke, and Wellesley in Massachusetts. A number of them also questioned Clarke's claims. Yet his book was accepted as the prevailing view, and it delayed the growth of higher education for women for at least a decade.[2]

As more and more women, and a growing number of men of color, chose to enter the sciences, American scientific institutions began to professionalize, a shift that made it more difficult to pursue science as an amateur. To be a respected scientist, one usually needed to earn a doctoral degree. This requirement gave men in powerful positions yet another opportunity to reject women from scientific disciplines, using the excuse that they needed to encourage higher standards. Even women with degrees met strong resistance when they tried to move up in the scientific hierarchy.[3]

The predominant European-American worldview about women shaped the thinking of major male scientists, among them Charles Darwin, who described women as physically and intellectually inferior in *On the Origin of Species* (1859). A decade later, in *The Descent of Man, and Selection in Relation to Sex* (1871), he emphasized male competition as the element of sexual selection that drives the transmission of certain traits from generation to generation. Though he acknowledged that female choice played a role in this process, he minimized its impact, describing the female influence on selection as "comparatively passive." His conception of males as active and females as passive perfectly reflected traditional Victorian assumptions about "the weaker sex." He wrote, "The chief distinction in the intellectual powers of the two sexes is shown by man attaining to a higher eminence in whatever he takes up, than woman can attain, whether requiring deep thought, reason, or imagination, or merely the use of the senses and hands." As a result, "man has ultimately become superior to woman." Wherever he looked, Darwin found evidence for the assumption that social inequality reflected biological inequality.[4]

Once again, there were women who questioned such conclusions. One was Caroline Kennard, an educator and amateur science enthusiast from Brookline, Massachusetts, a Boston suburb. After attending an 1881 discussion about the intellectual inferiority of women based on physical attributes, in which the speaker had cited Darwin as a "scientific authority," she wrote a letter to Darwin. As an admirer and a "believer in scientific discoveries and revelations," she was certain that Darwin had been misrepresented by the speaker, and she asked him specifically about his position regarding the inferiority of women to men.[5]

In his reply, Darwin told Kennard that women, when compared to men, had superior moral qualities but inferior intellectual qualities. Based on the laws of inheritance, it was unlikely that they could ever "become the intellectual equals of man." Kennard's return letter reflected her shock. She argued that women had the same capabilities as men but had not been given the same environmental or educational opportunities to develop their intellect. Nonetheless, Darwin's views were seen as gospel, and they would deeply shape the field of biology over the next 150 years.

In the 1880s, when Kennard was writing to Darwin, women were almost invisible in the sciences in the United States, but they were there all the same. Their apparent absence was not due to lack of merit or contribution. Rather, their presence was disguised and concealed. In these years, a growing number of white, middle-class women were gaining access to higher education and thus employment. At the same time, formerly enslaved Black women had few educational opportunities and were distanced from universities, colleges, and laboratories. Yet even as white women's educational levels increased and their roles outside the home expanded, they were expected to limit their work to a narrow range of activities that were stereotyped as delicate, feminine, and emotional. The stereotype of science was just the opposite: tough, masculine and unemotional. Thus, a woman scientist at the turn of the twentieth century was seen as a contradiction in terms. The historian Margaret Rossiter writes, "As scientists they were atypical women; as women they were unusual scientists."[6]

Ellen Swallow Richards is an interesting example. In 1879, she applied to the Massachusetts Institute of Technology (MIT), hoping to study for a graduate degree in chemistry, and was turned down solely on the grounds that the department did not want its first graduate degree to go to a woman. Such arguments snowballed over time: even when women were not the first to apply to a degree program, they were told that there was no precedent for admitting them. This circular logic seemed to close the issue forever. Richards persisted, however, and was eventually accepted at MIT as a tuition-free *special student*, a label that concealed the fact that she was officially enrolled, and allowed the institution to avoid granting her a degree.[7]

Despite these barriers, a growing number of women were serving as science professors at women's colleges, and more and more women were studying science. In 1876 and 1880, the Association for the Advancement of Women surveyed women scientists from across the country, with the goal of assessing their interests and obstacles. The results showed that most of them were isolated from the universities where major laboratory research was funded. Women remained outside the scientific community, looking in.[8]

Women scientists worked hard to overcome this separation. Yet even though some women pressed universities and employers to offer them graduate degrees and full equality, others were reluctant to be confrontational and tried to use existing sexual stereotypes to create new areas of women's work in the sciences. In the 1890s, for instance, Ellen Swallow Richards, who had faced so many barriers at MIT in the chemistry department, helped establish the field of home economics as a place for herself and other women.

By the early twentieth century, the segregation of American women scientists was firmly entrenched. Their jobs were limited largely to faculty positions at women's colleges and coeducational schools and to so-called "women's work" at research and government institutions. Some universities and colleges were willing to educate women in science, but they did not want to employ them except, occasionally, in junior positions. As the decades went by, most people took this employment pattern for granted. But the situation did not improve, and eventually even traditionally woman-friendly faculty positions began to be phased out because those colleges wanted to "professionalize"—which meant hiring more men.[9]

Black women faced even higher barriers. Until after the Civil War, most were enslaved, and those who were not had little access to colleges, universities, or laboratories. Still, by 1940, five African American women had completed their PhDs in science. Roger Arliner Young was one of them. After graduating from Howard University in 1924, she published an article about a paramecium in the prestigious journal *Science*, and in 1926, she completed her master's degree at the University of Chicago. In the late 1920s and 1930s, she worked at the Marine Biological Laboratory in Woods Hole, Massachusetts, and she earned her doctorate in zoology in 1940.[10]

Most people in academia assumed that women were not equal to men in science. Maria Mitchell and Marie Curie were seen as exceptions, but Young and many others were marginalized and largely invisible. The academy presented itself as a meritocracy, but women and people of color were in a perpetual struggle to conduct research and earn respect. A few might succeed, but those in each new generation had to prove their worth again.[11] The sexual segregation of the scientific workforce often obscured and devalued the work of women scientists, presenting it as merely technical. In contrast, as Naomi Oreskes argues, men's scientific work was often presented as heroic.[12]

Women scientists rarely received funding or promotions. When they did, their advancements were often treated as personal gifts from powerful colleagues rather than as rewards for good work. Black and white women, all people of color, and the working class were usually offered only marginal positions, the sort that were seen as most suitable for inferiors. This meant that women scientists were generally dependent on the goodwill of their male colleagues. Because many women worried that any feminist organizing would make matters worse, most chose to accept the situation and endure. But this also meant that they were isolated from one another.[13]

Between the 1940s and the 1960s, thousands of women worked in the sciences in the United States, but few of them were listed in the important biographical reference *American Men of Science*. (The title of the book speaks for itself.) Instead, they continued to hold junior positions and were seldom offered senior jobs, despite their qualifications and experience. This was true even for women who were studying at elite universities such as Harvard and MIT. In Boston, they worked in one of the largest centers of scientific research in the world, yet they continued to struggle for support.[14]

Ruth Hubbard, a biologist studying vision, and Rita Arditti, a geneticist trying to understand the lac gene, were determined to make their careers in science. Before the women's movement, they were seen as oddities or exceptions and were not fully recognized by their peers. In 1969, one of them would lead a protest at a science conference, and the other would look on in disbelief. But for decades before that moment, they were alone in their labs.

CHAPTER 2

INDIVIDUALS PERSIST

In the fall of 1941, when Ruth Hoffman Hubbard arrived at Radcliffe College, she was immediately faced with a hurdle: teachers who didn't want to teach her. Traditionally, Radcliffe was dependent on professors from Harvard, who would first lecture to large groups of men and then would walk over to Radcliffe and give the same lecture to women students, always a much smaller group. Many professors resented this system. In Hubbard's freshman chemistry course, the teacher, a recent Dartmouth graduate, treated the women students with disdain. He clearly would rather have been doing almost anything else.[1]

At this time, most of the women listed in *American Men of Science* had received their undergraduate degrees from women's colleges such as Mount Holyoke, Barnard, Smith, Vassar, and Wellesley. Radcliffe was at the bottom of the list. As Hubbard later noted in an interview, Radcliffe proudly offered its students the privilege of being taught by "Harvard's Great Men," but students did not get the idea that they could possibly become great women.[2]

During the summer after her freshman year, Hubbard took an introductory physics class that was ostensibly co-ed, though she was one of only two women in a class of 350. At the time, she thought there were so few women in the course because women weren't good enough at physics. She had no feminist consciousness at all. On the whole, Ruth identified more with men. She was much closer to her father than to her mother, both of whom were doctors. She knew where the power lay in her family and in the world, and she wanted to be part of that. She regarded women as second-class citizens and wanted to make damn sure she wasn't second-class herself.[3]

In fact, she'd been brought up among professional women. Ruth's mother and all of her mother's friends in 1930s-era Vienna were doctors. She had been raised in a middle-class, Jewish household that could afford to pay for childcare and housekeeping; it had never occurred to her that her mother should stay at home and do these tasks. Her mother was disparaging of women who didn't work, so Ruth assumed that she would follow in her mother's footsteps and earn an income. She knew she was expected to become an educated professional, unlike most middle-class white women in the United States, who assumed that they would not.[4]

On March 12, 1938, the night that the Nazis invaded Austria, Ruth's father received a cable from a friend in New York City urgently advising the Hoffman family to leave. A few days later, Ruth and her younger brother stood at their apartment window and watched the Nazi army march into Vienna. Within three months of Hitler's invasion, the family fled the country.[5]

After spending a month in New York City, the Hoffmans traveled to Boston. When they first arrived, the family lived in a cramped apartment on the Riverway in South Boston, and Ruth walked to the Girls' Latin School. She could understand English but could not speak it well; and as a result, her teacher told her that she shouldn't be there. Ruth hated the place.[6]

Meanwhile, the news from Europe was horrifying. From Boston, the Hoffmans heard about the Kristallnacht pogrom in Vienna on the night of November 9, 1938, when all of the city's synagogues were destroyed and most of the Jewish shops along the Ringstrasse were plundered and demolished. More than 6,000 Jews were arrested that night. Had the Hoffmans stayed, they would likely have died. As it was, they suffered vicariously, later learning that one of Ruth's favorite cousins had been killed in Auschwitz.[7]

Eventually, the family settled in Brookline, a suburb of Boston, and Ruth, withdrawn but self-reliant, immersed herself in books at the local public library. She had no interest in American teenage life—boys, clothes, dating. Instead, she read popular science books by Albert Einstein, Ernest Rutherford, and Niels Bohr. She was interested in physical explanations of the universe. By the end of high school, she still wasn't sure what she would study; but as the daughter of doctors, she thought it might be medicine.[8]

During her freshman year at Radcliffe, Ruth Hoffman met Frank Hubbard, a senior at Harvard who was studying English literature. His parents were from the Midwest, and he had been raised in Westchester, New York. The couple married in the middle of her sophomore year. "As American as apple pie" was how she later described her husband and her marriage. For the first time, she celebrated Christmas and Thanksgiving, and her new American name helped distinguish herself from her Old World parents. Even her voice started to change, sounding less Viennese and more Bostonian. She'd undergone an enormous change—from a fourteen-year-old refugee outcast to an eighteen-year-old married American woman. But as her father warned, "Getting married isn't going to protect you against the world."[9]

In the fall of her third year at Radcliffe, Hubbard, now a biochemistry major, started a research project on rat nutrition in Professor George Wald's lab in the biology department at Harvard. She enjoyed the environment both scientifically and socially. Lab personnel ate lunch together and talked about all kinds of topics. Using her research findings, she wrote her thesis about the physiology of hunger and activity. She was beginning to feel like a scientist.[10]

A few months before she graduated in 1944, Hubbard was naturalized as a U.S. citizen. She was nineteen years old, and she'd decided to stay on at Harvard, in part because Wald had offered her a job in his lab, one relating to vision measurement and the eye's sensitivity to infrared light. It was a military-related project: the U.S. Army needed infrared filters and devices that would allow troops to maneuver in the dark. Hubbard knew that Wald could be domineering, but he reacted well to people who asserted themselves, which gave Hubbard confidence. She appreciated that the research was theoretical and practical, and she enjoyed the experiments and working with machine tools.[11]

Thanks to record growth in government science budgets, young male scientists saw many new opportunities open up for them after World War II. The situation was different for women scientists, who continued to be marginalized and underused. Moreover, younger women, many of whom had been working during the war and who might have chosen to stay employed or seek higher education in the sciences, were now expected to stay home with their children.[12] Still, many scientific women persisted, even under adverse conditions.

In 1946, Harvard Medical School began accepting women applicants, so Hubbard had to make up her mind about whether or not to apply. In the end, she decided to pursue her doctorate in scientific research. She didn't want to relinquish the life of the lab, the focus on intellectual curiosity, the sharing.[13] Still, a junior faculty member told her that she should go to a graduate school where she had a chance of becoming a faculty member. Hubbard didn't really understand what he meant, however, and she stayed on at Radcliffe.[14] She did begin to spread her wings, though, spending a year working in a lab in London, delivering a paper at the Biochemical Society Meeting and at the First International Congress of Biochemistry in Cambridge, England. From then on, she continued to write and deliver papers, building her expertise and boosting her confidence.[15]

Hubbard had been working hard to become ultra-American, but when she and Frank traveled in France, she began to think of herself as European again. She realized that she could relax in Europe, feel at home there in a way that she had never felt in the United States. She also recognized that her marriage wasn't going to last. After spending time on their own, Ruth and Frank officially separated.[16]

Hubbard completed her PhD in 1950 but never gave up her work in the lab. Science was a combination of sleuthing and luck, she'd decided. You asked a question hoping for an answer, and then worked on it for a while. If your work didn't give you a clear outcome, then you had to stop and ask another question.[17]

When she'd started working in Wald's lab, no one yet understood the chemical reactions in the eye that made sight possible. Decades earlier, however, Wald had found that vitamin A was an essential part of the molecules of visual pigment in the retina. If an eye lacked vitamin A, it couldn't make the pigment necessary for sight. Scientists knew there was a connection between diet and vision but didn't have a clear understanding of the underlying chemical mechanism. Part of Hubbard's work was to try to understand the chemical reactions.

For her PhD, she worked on the question of how vitamin A gets converted to the chemical compounds required for sight. She found that the shape of the vitamin A molecule was important in connecting to the protein and forming the visual pigment. She also found that light had an impact on the shape of the molecule and that it took time for the molecular shape to change. This is why it takes a while for our eyes to adjust when we go from light into darkness. Over the next few years, Hubbard worked to further understand this chemical reaction in the eye. Eventually, she was awarded a Guggenheim fellowship, which allowed her to travel to the Carlsberg labs in Copenhagen, the center of protein chemistry.[18]

Hubbard loved the environment of working in a laboratory. Because of her light-sensitive work, she often sat at the lab bench bathed in red, as if in a photographic darkroom. Hair tucked behind her ears, she would bend over a cow's eyeball. Inside that eye, at the back, was the natural equivalent of photographic film—the retina. In her drive to understand how vision worked, she needed to look more closely at the proteins in the cells of the retina.

Rod cells are photoreceptors in eyes, and they contain pigments made of proteins. When exposed to light, these visual pigments absorb the light and create nerve impulses that are transmitted to our brain, where the light creates the sensation of seeing—vision. Hubbard wondered how our biological camera worked. What exactly was the biochemistry of vision? She was particularly interested in rhodopsin, one of the pigments in the rod cells.

With fine scissors, she would cut through the vitreous gel of the eyeball. Next, using miniature forceps, she would carefully lift off the cornea and the lens and put them to one side. After immersing the eye in a saltwater solution, she would reach in with a tiny spatula to loosen the retina, a transparent veil, and watch it float away from the eye into the solution.[19]

After she had gathered enough retinas, Hubbard could begin her biochemical analysis of the rod cells. To conduct these studies, she would pipette small volumes of samples from test tube to test tube with careful precision and, using spectrophotometers, study the way in which the sample interacted with light.[20]

For the bulk of her scientific career, Hubbard was focused on the question, How does this camera work? Her list of scientific publications was impressive, and her work on vertebrate and invertebrate visual pigments set the experiment procedures for much of the research in that area. A young colleague, Paul Brown, said that Hubbard's papers were landmark contributions to the knowledge of visual pigments and their function in the eye. She was invited to speak at many symposia and lectures in the United States and internationally. During the 1950s, whenever a young postdoctoral fellow arrived at Wald's lab, Hubbard was the one to help him learn how to cut open an eye, interpret an experiment, and draw a figure for a paper. Yet despite her skills and achievements, she never advanced beyond the position of research associate. No one offered her a professorship at Harvard or anywhere else, even though Wald once told a *Boston Globe* reporter that she was "the best graduate student I ever had."[21]

Hubbard liked her work in Wald's lab, and she was productive. She doesn't seem to have ever applied for faculty positions at other institutions. Perhaps she was waiting for an offer that never arrived; perhaps she knew it was unlikely that she would be considered for other positions. By the time she was in her early thirties, she had spent almost fifteen years in Wald's lab. She was not building an individual career path. But in the 1950s, as a woman, she was socialized to think that she didn't have to be too concerned about that.[22]

In 1957, Evelyn Fox Keller came to Harvard full of enthusiasm. It was her first day of graduate school in the physics department at Harvard, and she was a long way from her beginnings. She'd grown up in New York City, the daughter of Russian émigrés who worked long hours in a delicatessen to support their family. They had supported Keller while she completed her degree in physics at Brandeis, a new university outside of Boston founded by the Jewish community, and they were proud that she had been accepted at Harvard for her PhD.

At twenty years old—twelve years younger than Ruth Hubbard—Keller was tall and slim. She wore her dark hair long. But when she sat down to talk with her adviser, the first thing he said was that she couldn't take the course she wanted. The second thing was that she shouldn't concern herself with quantum mechanics. As he explained that her previous work at Brandeis had no meaning at Harvard, Keller could sense both his arrogance and his class bias. And he didn't like her outfits—mainly skirts and sweaters with a bohemian flair, very little makeup. Instead of focusing on her physics career, her adviser wanted to teach Keller how to become a suitable date. He even assigned a senior male graduate student to teach her how to dress.[23]

One day, after a seminar at MIT, a fellow student offered Keller a ride back to Harvard, which she gladly accepted. As he drove, he asked her how her courses were going. Assuming that he was sincere, she let her guard down and told him how terrible her experience had been in the Harvard physics department. She felt invisible, she said. "No one talks to me." But when she glanced over at the driver, she noticed a look of acute discomfort on his face, as if she had just committed a terrible faux pas. She had said too much. When she got out of the car, she felt great shame.[24]

As a child, Evelyn had had no such anxiety about her place in the scientific world. She was always encouraged by her much older siblings: Frances, who became a sociologist, and Maurice, who became a biologist and would load her down with science books.[25] Though her parents didn't want her to leave home, they eventually let her go to Brandeis, and she found support there from several Jewish scientists. When Sam Schweber, her physics professor, started writing on the board, Keller thought his equations looked like a fascinating alien script or an obscure religious text. She copied everything down and spent her weekends trying to figure out what the new language meant, taking books out of the library to help her decipher the code. She loved that the equations helped her to think about how the natural world worked. She loved asking questions about why space was three-dimensional and thinking of the answers in theoretical terms. Keller took a course in theoretical physics and wrote her thesis about the Nobel Prize–winning physicist Richard Feynman and his Lagrangian formula of quantum mechanics, which helped to explain energy and motion. She had found a new passion.[26]

But there were warnings of trouble ahead. At a party, Keller told the famous scientist Leo Szilard, a family friend, about her plans to become a theoretical physicist. He told her this wasn't possible because women didn't do physics. Not yet twenty years old, Keller had heard such comments before, but she didn't think they applied to her. She took his response as a challenge.[27]

Through physics Keller had been introduced to the concept of "great men with great minds," a notion that was thriving in the 1950s. Einstein was one, Feynman another, Szilard a third. She was absorbing the idea that only men could have great minds.[28] Still, she wasn't prepared for Harvard. In an essay written decades later, she described her two years in the physics department as two years of insult. Because she didn't have a political or feminist framework for comprehending the situation, she responded with rage. When she found that rage didn't help, she became depressed. Friends and family kept asking what was wrong with her, as if her misery were somehow her fault.[29]

Most painful was the isolation. Briefly, Keller made friends with another graduate student in her class, a man who had been a Harvard undergraduate. They sat together in the classroom and sometimes walked to the dining hall together. They met in the library and spent hours talking—she explaining the physics problem sets to him, he explaining the workings of Harvard to her. But one day her friend decided that being associated with her was a risk to his reputation, and he asked her to stay away. After that, she was totally alone. She arrived early for lectures and took a seat in the classroom, watching as it filled up with students. The sea of empty seats around her only added to her sense of isolation. Keller worked hard, did well in her courses and exams, and sought affirmation from her male colleagues without success. Her primary goal became to *appear* to be surviving.[30]

When Keller handed in good work, she was suspected of plagiarism. Her committee chairman didn't show up for her oral exams. Two months later, he admitted that he owed her an apology, his excuse being that he had taken two sleeping pills and overslept. But the exam had been at 2 p.m., so she wasn't convinced. Nonetheless, despite the discouragement, she made it through her difficult first year. [31]

That summer, Keller's brother Maurice invited her to join him and his family at Cold Spring Harbor, New York, a leading research center for biology and genetics. Maurice was planning to take a summer course in phage genetics (viruses that affect bacteria, a hot topic among molecular biologists), and he thought his sister would enjoy the environment. She did. She found the scientists friendly and welcoming, and the atmosphere was thick with excitement in the wake of James Watson and Francis Crick's recent paper on the structure of DNA. Soon Keller was working in the laboratory alongside Matthew Meselson and other biologists.

During her stay, Keller often saw the famed maize geneticist Barbara McClintock taking solitary walks in the woods and along the beach. The two women briefly worked in the same building, but Keller never talked to McClintock or visited her lab. Somehow McClintock held herself apart from the hustle and bustle of the summer school and the molecular biologists. At the time, Keller wasn't curious about her, but that would eventually change. [32]

For Keller, the summer in Cold Spring Harbor was a lifeline, giving her not only the strength to complete her PhD in physics but also the encouragement to write a thesis that incorporated molecular biology. After finishing her degree in 1963, she moved to New York City and began to teach physics at Cornell University Medical College. She married Joe Keller, a mathematician, and they had two children. She was hopeful that her life in science would continue to thrive.

Born in Buenos Aires, Argentina, Rita Arditti had long wanted to go to the United States and work in a laboratory. So as Evelyn Fox Keller was completing her PhD, Rita Arditti was moving to Boston. It was the thing to do, she later recalled, if you were a serious biologist. On first arriving, she took up a postdoctoral fellowship in biochemistry at Brandeis, where Keller had studied, and found a place to live nearby. [33]

Small and short of stature, with large, dark-rimmed glasses, Arditti was a dedicated lab worker and put in long hours. One Saturday morning, as she was gathering supplies in the cold room, she unexpectedly ran into the head of the department. He seemed surprised to see her and asked her why she wasn't at home with her kid. At first Arditti thought he was being kind, suggesting that she didn't have to work on the weekend. But as he walked away, she felt that a different message was being conveyed: you don't belong here.[34]

Rita Arditti grew up in a Sephardic Jewish family in the 1940s. During the years of Juan Peron's rule, the national university in Buenos Aires was frequently closed. So in 1952, when Arditti was eighteen, she came to the United States to attend Barnard College in New York City for a year. There she met Mario Muchnik, another Argentinian, who encouraged her to study science. Unhappy to live in a dorm for international graduate students instead of with other undergraduates, Rita returned to Argentina with Mario but found the universities in chaos. As a couple they decided to move to Italy and attend the University of Rome, where Rita studied biology and Mario studied physics.[35]

Arditti was attracted to the rigor of biology. She was in Rome when Watson and Crick announced that they had pinpointed the double helix structure of DNA, news that she found incredibly exciting. She loved genetics because it was elegant and clear, and she believed she could contribute to this dynamic field. In the lab, she immersed herself in the genetics of microorganisms and labored harder than ever before, primarily on experiments involving phages. She also began publishing journal articles in *Genetics* and *Virology*.[36]

At the University of Rome, Arditti was one of only a few women doing genetic research. She felt hostility from some of her male teachers and colleagues, but she dismissed this as personality problems. She never thought it had anything to do with her being a woman.[37] However, her views shifted after she married Muchnik. When their son was born in late 1960, Arditti began to struggle with the competing needs of her marriage, her child, and her research. In the lab, she noticed the difficulties facing women scientists who were married to men in the same field. Regardless of their ability, the women held lesser positions and often had to accept a precarious work situation in which their research depended on their marriage. These women did the day-to-day work in the lab that was necessary to produce a solid piece of research, yet the credit would invariably go to their husbands. At first, Arditti thought she was safe from such treatment because her husband was in a different field. It was only much later that she reflected that all women were vulnerable to this kind of exploitation.[38]

When her son was eight months old, Arditti left her husband. She was twenty-six, had completed her doctorate, and had begun a postdoctoral fellowship in Naples at the International Institute of Genetics and Biophysics. The constant hostility she faced in the laboratory—all of her bosses were men; all of the laboratory heads were men—was taking its toll, and her excitement over her research was waning. The situation had become increasingly uncomfortable, and she had no peers to talk to. All of the women she knew were either technicians or students; none were PhD scientists like herself.[39]

In Naples, Arditti met Jon Beckwith, a geneticist from Harvard Medical School, and she and her son became friends with him and his family.[40] When she made the move to Brandeis, they reconnected. At Brandeis, Arditti was focusing her research on mutations in the promoter region of the lac operon in the bacterium E. coli, and she had published numerous articles about her findings in the *Journal of Molecular Biology*, the *Journal of Bacteriology*, and elsewhere. One day, as she was talking to Beckwith about her interests, he offered her a post as a research associate in his lab at Harvard. Arditti accepted, and she and Federico made the move from Waltham to Cambridge.[41]

At Harvard, Arditti and her colleagues were trying to understand how E. coli's production of lactose is controlled at the genetic level. They wondered if certain genes could be turned on and off by the involvement of repressor molecules. In the late 1960s, this was ground-level work; and Arditti's experiments, as well as those of her colleagues at the time, enabled many future discoveries in molecular genetics.[42]

To reach Beckwith's lab, Arditti had to travel through the halls and corridors of Harvard Medical School. With her son Federico's small hand in hers, she would gaze up at the portraits of the distinguished white men who had worked in these hallowed spaces. The building had a palatial feel, like Versailles. Few, if any, had needed to bring their small children to work with them, but Federico spent many afternoons in the lab. To keep him busy, she would give him a few scientific tchotchkes, and he would play with them on the floor while she leaned over a lab bench. Using a glass pipette, she would extract fluid from a test tube and deposit the correct amount in a series of petri dishes, then place the dishes carefully into a centrifuge that would spin the liquid at high speed and separate out various components that she needed for her research.[43]

Years later, Federico recalled one particular day at the medical school. He was sitting with his mother at the end of a very long table in a room outside the lab. In front of him was a chocolate cake with candles. Around him a group of men wearing glasses and pocket protectors were singing "Happy Birthday." He felt a little scared and wondered, Where are the other children? Where are the presents and the games? But in the 1960s, there was no child-care available at Harvard Medical School.[44]

But Arditti had made this choice with her eyes wide open. From a very young age she knew she did not want to be like her mother, whom she saw as dependent on her father, not in a position to "live a life of her own." Conversations at the dinner table could be tense. When Rita declared that she wanted to be a lawyer or a doctor, her father would make fun of her. To get away from his belittling comments, she chose to go to boarding school for three years of high school. She needed distance from her family's expectations.[45]

All of the women in Arditti's extended family were housewives and moth-ers; only one worked outside the home. This was typical in 1950s Argentina. An unmarried woman might work in an office or a bank, but she wasn't expected to continue her education after high school or to keep working after she got married. Every young woman in Arditti's family received the same message: "Men will take care of you." But what if they didn't get mar-ried? None of the girls had a clear idea of what life would be like for a single woman. Like her three sisters, Arditti married at twenty. But by pursuing science and a PhD, she still managed to evade her family's expectations.[46]

Still, in the mid-twentieth-century United States, most women in the sciences were not aware of their history of exclusion. They were not thinking about how institutions had created barriers, how societal norms were mak-ing it difficult for them to pursue their interests. They were not analyzing patriarchy or white supremacy. They were individuals, predominantly white, who were trying to overcome their challenges as best they could.

We know about Hubbard, Keller, and Arditti because their work was later recognized as important and because all of them produced oral histories. These interviews and recordings offer valuable information about their early careers through the lens of their later, more developed, political analysis. Such oral histories are less common among women of color in the sciences and are usually incomplete, especially for scientists who were active during this time period. This is one reason why white women's stories dominate the history of women in science.

Over the course of fifteen years together in the lab, Ruth Hubbard and George Wald developed a secret relationship. Finally, in 1958, they came out into the open and got married. This was the second marriage for both; Wald was eighteen years older than Hubbard and already had two sons. But Hubbard was ready to have children, and she assumed that her scientific reputation was secure by this point and that she could handle both motherhood and research. Their son Elijah was born in 1959, followed by their daughter Debbie in 1961.

Both Hubbard and Wald took it for granted that she would continue her research in the lab. One day she bumped into an acquaintance in Harvard Square, who asked if she were still working. "Yes," Ruth answered, puzzled. What an extraordinary question, she thought. Yet throughout the 1960s, she heard it regularly.[47] Still, neither Wald nor Hubbard challenged her role within the family. Given that he was a professor with students and committee meetings and she was a more flexible research associate, both assumed that she would take care of the children and manage their home.[48]

At this time, many women with PhDs were working as research associates in university labs and, like Hubbard, had limited upward mobility. Most needed to seek the approval of the lab head to apply for grants. But even with her heavy domestic responsibilities, Hubbard continued to perform her own research and write her own papers, including an influential one delivered at the 1965 Cold Spring Harbor Symposium on Sensory Receptors. She was also lucky in being able to negotiate her own grants with the National Eye Institute at the National Institutes of Health. Perhaps because of this, Harvard allowed her to continue her interesting work, and she was grateful. Nonetheless, she was part of Wald's lab and dependent on his infrastructure. At the same time, her presence was integral to his success. Though she didn't teach, mentor her own students, or control departmental decisions, she provided significant support to the lab and to her husband's students.[49]

In 1967, Hubbard and Wald shared the prestigious Paul Karrer Medal for their work, but this was only because Wald wrote to the organizers at the University of Zurich to ask if Hubbard could be recognized as well: "She has been my closest collaborator in much of the work I shall be talking about and is responsible for some of its most important aspects." However, the accompanying Paul Karrer Lecture was delivered by Wald alone, and until 2019 the medal's website did not include Hubbard's name as co-recipient. The correction was made only after the journalist Kate Zernike pointed out the error.[50]

Decades later, one of Hubbard's colleagues said that her work had a strong influence on the field of visual biochemistry. Yet in the late 1960s, her male contemporaries, many of whom were considerably younger and with fewer scientific accomplishments, had far more independence, rewards, and status. At the time, Hubbard didn't notice.[51]

On a crisp morning in October 1967, the phone rang at Hubbard and Wald's house on Lakeview Avenue in Cambridge. It was for Wald, and when he hung up, he said, "I've won the Nobel Prize." For a moment, the couple was in disbelief. They hadn't been working toward it or expecting it. Their son Elijah had already left for school, but they shared the news with six-year-old Debbie. "We're going to see the King of Sweden!" Debbie told the *Boston Globe* reporter who arrived later that day.[52]

That year Wald shared the Nobel Prize for Physiology or Medicine with the Finnish-Swedish scientist Ragnar Granit and Haldan Keffer Hartline from Rockefeller University in New York. Hubbard was thrilled for him. She rapidly started preparing for the festivities: found some green silk fabric, worked with a local dress designer, went hunting for matching shoes. The whole family flew to Stockholm, and Wald was treated like a film star for a week. The event's brilliantly lighted ballroom felt like a Hollywood set. As Debbie had predicted, the king of Sweden was there, and Elijah sat next to a princess.

After dinner, one of the men on the Nobel committee, the Swedish scientist Hugo Theorell walked over to where Hubbard was sitting, glamorous in her floor-length, off-the-shoulder gown. The champagne was flowing. A quartet was playing. In her elated mood, she thought for a moment he was inviting her to dance. Instead, he said, "You know you were kept out of the prize by the rule of three."[53] The music stopped. Hubbard stared at Theorell, trying to make sense of what he was saying, but he turned and walked away.

Theorell was referring to a rule in Alfred Nobel's will, which guides the presentation of Nobel Prizes: the committee cannot give a prize to more than three people at once. In other words, the founder did not want the prize money to be divided into portions that were too small.

But Hubbard was flabbergasted. Why had he told her this? Did he think she would feel honored that she had been considered for the award at all? Did he think she would be happy to hear she was fourth? The fact was that she wasn't honored or happy. She was bewildered. At the time, she didn't imagine that perhaps she also should have won or that sexism had been at play. But many years later, when she was looking back at the moment, she saw it as a dreadful incident. No one on the Nobel committee would ever have said such a thing to a man.[54]

There was a history to such situations, as Hubbard later realized when she learned about Rosalind Franklin's experiences. A brilliant young British physical chemist, Franklin contributed to the understanding of the structure of DNA in the early 1950s, with a photograph she took using X-ray diffraction. She died in 1958 at the age of thirty-seven, four years before James Watson, Francis Crick, and Maurice Wilkins received the Nobel for their DNA research. Had Franklin still been alive, the question of the rule of three might have arisen. More disturbingly, in 1968, Watson published *The Double Helix* in which he minimized Franklin's contribution and wrote that he had stolen her notes without her knowledge. Hubbard knew Watson, who was at Harvard between 1956 and 1976. Yet at the time of Watson and Crick's Nobel, she was not fully aware of Franklin's role.[55]

However, at least some contemporaries were aware of Hubbard's own unheralded role. After the couple returned from Stockholm, they were invited to a potluck dinner party, where the talk mainly revolved around the Vietnam War. Standing with her plate, Hubbard was chatting to friends when another person pulled her aside to ask, "Did you win George's Nobel Prize for him?"[56]

Hubbard had certainly played a large part in Wald's success. Yet after he won the Nobel, her reputation as a scientist changed dramatically. From that point onward, her research was seen as his work. Moreover, her record in the literature started to read differently. Slowly, Hubbard began to sense that all of her own contributions were being erased. Everything she had ever done was being described as Wald's work.[57]

Twenty-five years later, Margaret Rossiter, a historian of science gave this trend a name, "the Matilda Effect," and used Ruth Hubbard as an example. As Rossiter was able to show, it was common to subsume a woman scientist's work into the accomplishments of a better-known male colleague. Two years after Wald's Nobel, Hubbard herself was beginning to notice these discrepancies. And in 1969 her attitude about them was about to change.[58]

CHAPTER 3

EPIPHANIES AND DISRUPTIONS

The formal luncheon in the plush ballroom of the Sheraton Boston had just gotten underway when the doors burst open and twelve women marched inside. Some were dressed in bellbottoms, others in miniskirts. They were carrying posters, one of which read, "Female Scientists Do Not Escape Oppression." A small woman in large, dark-rimmed glasses shouted in an Argentinian accent, "Women in science are relegated to second-class status!"

It was the last week of December 1969, and they were interrupting a special lunch for women scientists, hosted by the scholastic honor society Sigma Delta Epsilon and taking place during the annual conference of the American Association for the Advancement of Science (AAAS). Quickly, the protesters got to work. Some moved from table to table, handing out mimeographed leaflets, while Rita Arditti scanned the room and prepared to speak. Although she held a doctorate in biology and was working in a genetics lab at Harvard Medical School, she was not on the program of presenters. Around her, women looked up from round tables covered in crisp white linen—forks poised, concern on their faces. Their attitude was cool and distant, but Arditti pressed on. "Women are oppressed economically and culturally," she declared. "They are trained for inferior roles and exploited as sex objects and consumers."[1]

Most of the diners turned away and continued to eat their lunch. But one, a woman with a long braid wrapped around her head (a nod to her Viennese upbringing), did not. Ruth Hubbard scanned the messages on the placards and listened to what Arditti was saying. But she was confused: she didn't understand why the protesters were lambasting Harvard for its stance on women. Now in her mid-forties, confident in her role at the lab, with a PhD and a string of prestigious publications, she had not, to this point, been unduly troubled by the position of women scientists. Still, the situation had caught her attention, and she picked up a leaflet and skimmed Arditti's words.

In the leaflet Arditti noted that the AAAS's stated goals were to further the work of scientists and increase public understanding of the importance of science. Neither of those objectives could be realized, however, so long as women in science were relegated to second-class status.[2] "While men are trained to develop 'logical' patterns of thought, women are encouraged to be 'intuitive.' Math and science are seen as male prerogatives." According to Arditti, women scientists were limited by "being placed in subordinate positions": they were rarely given their own labs or first authorship on papers, and they were paid less than their male colleagues for equal work. Activists were urging AAAS to adopt a resolution to end sex discrimination in science.[3] Afterward, one of the protesters, a student, told a *Boston Globe* reporter that the women at the luncheon were not courteous, but "hopefully it got them thinking and that's a big step."[4]

Hubbard slipped the leaflet into her bag. The protest had given her a jolt, though she did not immediately connect the dots. Arditti's resolutions struck her as odd. Hubbard didn't understand what they meant or how any of it applied to her. Yes, she was a woman. Yes, she was a scientist. Harvard had allowed her to work in her husband's lab as a research associate for close to twenty-five years, so where was the discrimination?[5]

That week, a nor'easter was blowing through Boston, bringing frigid temperatures and hurricane-force winds. Nonetheless, more than 10,000 scientists had braved the cold to attend the 136th annual meeting of AAAS, the largest professional science association in the United States. Buzz Aldrin, an astronaut who had taken part in the Apollo 11 mission to the moon, was on a panel titled "Science and Man in Space." Several White House science advisers who had served under presidents Eisenhower, Kennedy, and Johnson were attending, as was the anthropologist Margaret Mead. Boston's public television station, WGBH, was preparing to air interviews and live broadcasts. And for weeks beforehand, a group of students and scientists had been planning to disrupt the entire event.[6]

To understand how Rita Arditti came to lead the protest that disrupted the women scientists' lunch that day, we must first pause to consider what had been happening in Boston, in the United States, and to Arditti personally during the previous two years.

The year 1968 had been filled with upheaval. Dissatisfaction with the Vietnam War was growing. Young people were questioning not only the conflict but also authority and mainstream politics. Martin Luther King Jr., who had spoken out against the war, was assassinated in March. Three months later, Robert F. Kennedy, a popular Democratic candidate for president, was assassinated at a campaign event in Los Angeles. Shortly afterward, Arditti and many others sat by their televisions watching the police assault antiwar protestors at the Democratic National Convention in Chicago.

Amid this chaos, the women's movement in the United States was growing, and its reach was spreading. So when a friend invited Arditti to an informal consciousness-raising gathering about women's lives, choices, and work, she was interested. Being a woman was political, her friend said. Although Arditti had no idea what this meant, she liked and respected her friend, so she went along.

Arditti did not record who attended the meeting or at whose house it was held; she just noted that about twenty women were present, all sitting on the living-room floor. She was the only foreigner in the group, and she did not immediately feel at home. As she listened to the others talk about their job and home concerns, she wondered if these feminist issues were relevant to American women only. Perhaps she didn't fit in because she was from Argentina or from a Sephardic family. She didn't see how their problems related to her own life.[7] She was shocked when one woman started talking about sex, another about orgasm. How could they discuss such personal issues with people they hardly knew, share intimate details in such a non-intimate way? Mostly the meeting made her uncomfortable; and when she got home that night, she resolved never to attend another one. Nearly a year passed before she gave the issues it had raised any further thought.[8]

The anti-Vietnam protests, not the women's movement, were what pushed Arditti to see science as a conservative force in society. During 1969, the war and the swelling antiwar movement, especially on college campuses across the country, captured her attention. She was developing serious concerns about militarism in science, and she was not alone. On March 4, 1969, a group of engineers, scientists, and students led a major protest at MIT to stop all war-related research on campus. In an extraordinary mass meeting known as "Scientists Strike for Peace," they protested the institute's complicity. George Wald, Nobel laureate and Ruth Hubbard's husband, delivered a speech that was widely reprinted: the *Boston Globe* alone distributed 87,000 copies of it. The event rocketed Wald into the public realm as a peace activist.[9]

As many young American men were being drafted into military service, most university students were exempt because they were enrolled in school. Nevertheless, the U.S. military and the Reserve Officers' Training Corps (ROTC) were a visible recruitment presence on college campuses, Harvard among them. Students for a Democratic Society (SDS), a campus-based group with branches throughout the country, protested against both the war and the ROTC. On April 8, 1969, they ratcheted up their activism at Harvard, posting a list of demands on President Nathan Pusey's door. The following day students forced their way into University Hall in Harvard Yard and occupied the administrative offices throughout the night. The next morning Pusey asked local police and state troopers to remove the students. Wearing helmets and carrying batons, the police marched onto campus and violently removed more than two hundred of them. The air filled with screaming as students were trampled. Afterward, the University Hall steps were stained with blood. The shock of that day ignited a student strike: thousands joined the uprising against the war and the university.[10]

It is not clear if Arditti attended these protests at MIT and Harvard. Nonetheless, both events had a major impact on her. They were so close by: Harvard Medical School, where she worked, was located just across the Charles River from Harvard Yard. She began to engage with other concerned scientists and in early 1969 became one of the first members of a fledgling organization known as Science for the People. The group's formation did not happen at a single meeting or in one place but over time as several different groups from around the country came together. First, a group of physicists introduced an antiwar resolution at the American Physical Society convention. Then a group in California called Scientists for Social and Political Action (SSPA) began to hold meetings and recruit members. Before long, the name changed to SESPA (to include engineers) and produced a nationally distributed newsletter. Printed with the slogan "Science for the People," it became a communication channel for various study groups and caucuses around the country. Toward the end of 1969, they began to gel into a single organization.[11] Finally, in early 1970, the newsletter became a bi-monthly magazine called *Science for the People*, and the national organization assumed the same name. The magazine was published for more than twenty years.[12]

At first, Science for the People's primary concern was the war in Vietnam, and members were vocal in speaking out against the militarization of science. More broadly, however, the organization supported the idea that science should support the needs of the working class, the oppressed, and humanity in general rather than be developed in the interests of the elite and the military.[13] In the fall of 1969, the fledgling group began to plan protests at the December AAAS conference in Boston.[14]

In the trajectory of Arditti's activism, the war came first. Her concerns about women's equality in science were secondary. But over the course of 1969, she began to understand that science had created and was continuing to reinforce stereotypes and images of women that were conservative and traditional. The frank discussions at the consciousness-raising meeting she'd attended in 1968 had planted a seed in her mind, just as her disruption of the women's luncheon would plant a seed in Hubbard's.[15]

Arditti started to think about her life as compared to the lives of the men in her field. She started to look at the scientific world with a different eye. She noted that there were very few women in positions of authority and, for the first time, wondered if some of the difficult dynamics she was experiencing in the laboratory were linked to the way in which men viewed women.[16]

Men with equivalent education, intelligence, and enthusiasm were all moving up in their careers. Why wasn't she? She was working very hard. She was producing results, giving papers, yet she wasn't advancing in her career. For years she had felt like she was existing in a fog, but she'd never before been able to put her finger on the cause.[17]

Arditti started talking about these matters with other women scientists. She began to bristle at the patronizing attitudes of male colleagues. She felt a lack of respect, support, and recognition. The sensation was familiar. Being a woman scientist in America felt like being a Sephardic Jew in Argentina.[18]

When she was a child in Buenos Aires, Arditti hadn't known that the rest of the world wasn't Jewish. She and her two sisters had loved to visit her three cousins and her Aunt Selma for Pesach. The table would be laid with

Sephardic dishes such as okra in tomato sauce and borekas with eggplant. The six little girls would collapse with laughter when the men read religious books with their hats on.

Then came Catholic elementary school and Rita's gradual realization that she lived in a Catholic country. She was one of only a few Sephardic Jews at the school, which had a larger group of Ashkenazi children, but they, too, were a minority. The other students didn't believe she was Jewish because she didn't speak Yiddish and her parents were from Turkey. Perhaps, also, their assumptions were influenced by her Italian name and her "Mediterranean looks." Catholic students made antisemitic comments in front of her, assuming she was Catholic. That experience convinced Arditti to always let people know that she was Jewish. Gradually, she recognized that to be a Sephardic Jew in Argentina was to be "a minority within a minority"—to be invisible. Now, as an immigrant, Arditti felt like a foreigner in the United States. As a woman scientist in the academy, she felt like a minority within a minority. She was coming to believe that her experiences in Argentina had made her "a socially aware person."[19]

In the late 1960s, Arditti was still working in Jon Beckwith's lab at Harvard. She began noticing that he was making great efforts to get other men into faculty positions but was not paying the same attention to her and her career. She was by no means the only woman in this situation. At the age of thirty-five, she seemed to be at a dead end. It was possible, she thought, that she would stay in the lab forever, doing interesting work as someone else—a man—continued to get the credit.[20]

With the December AAAS conference looming, Arditti was one of several scientists who argued for introducing three resolutions to the organization's ruling body. The first resolution would call on the society to demand the withdrawal of U.S. troops from Vietnam. The second would denounce repression against the Black Panthers, labeling it "systematic subjugation," and would call on scientists to fight against the use of technology and scientific knowledge in support of law enforcement's "war against blacks." The third asked the scientific community to implement an eight-point program to make sure that women were not discriminated against in science. A small group of feminists drafted the declaration, and Arditti put it squarely in the forefront when she and her colleagues—whom the *Boston Globe* described as "an intellectual horde of unhappy women"—invaded the luncheon.[21]

All three resolutions were controversial in themselves. But, in addition, many scientists didn't like the idea of discussing politics at a scientific conference. "Knowledge is neutral," said Walter Roberts, the outgoing president of AAAS at one of the conference plenaries. Protesters groaned in response as he tried to make the case that science and technology were neutral fields outside of the political realm. This ran exactly counter to the beliefs of Science for the People activists. They wanted to demonstrate that science was infused with the sociopolitical issues of the day.[22]

The actions at the AAAS conference were a major turning point for the fledgling organization. It was there that Science for the People first introduced its iconic fist-and-flask logo: one hand making a power fist, the other holding a lab beaker. Protestors handed out hundreds of buttons at the conference, inserting both their name and their logo into this large gathering of scientists.[23]

On the last day of the conference, more than 150 AAAS council delegates, representing more than 120,000 association members, filed into the carpeted hotel ballroom and took seats on folding chairs. The council president stood before them at a podium, waiting to start the proceedings. Behind a velvet rope in the visitors' gallery sat about thirty activists from Science for the People. Although they had gathered several hundred petition signatures, they had not convinced the AAAS to include their three resolutions on the official agenda. Thus, they were present only as observers, relying on the hope that one of the council representatives would raise their resolutions from the floor.[24]

The proceedings began, and the council elected Mina Rees as the first woman president of AAAS. Rees was an internationally renowned mathematician and had been involved with the development of the first high-speed computers. According to the *Boston Globe*, she must have missed the message that "little girls are supposed to be incapable of abstract thought." In fact, however, Rees was full of contradictions. In her acceptance speech, she told the plenary that many women were content with homemaking and child rearing and didn't want to be professionals. She advised women who wanted to be scientists to "marry the right husband." She did acknowledge the young female "liberation-type groups" at the conference and acknowledged that she, personally, had experienced numerous incidents of exclusion during her career. She told the plenary that the protests were important because no institution would change unless it were pushed.[25]

After moving through several items on the agenda, the council chair then asked for new business. The ballroom remained quiet. No delegate offered anything from the floor. "Have none of you the courage to face the real issue?" yelled a man from the gallery. "Who do you represent, anyway?" a young woman shouted. The delegates sat stiffly in their seats, facing the podium. The council chair hastily adjourned the session, but not quite in time.[26] Swinging their legs over the velvet rope, students rushed forward from the gallery onto the ballroom floor. One picked up the podium mic and announced, "The first people's council of the AAAS has convened." At this point, most of the council delegates walked out. At the podium, a student read the resolutions. With a voice vote, Science for the People elected to withdraw all troops from Vietnam, put an end to the repression of Black people, and call a halt to all discrimination against women in science.[27]

This symbolic gesture was the beginning of Rita Arditti's work with Science for the People. In her personal capacity, she would go on to question the field she had embraced for her entire career. As her determination strengthened, she would also begin to critique the way in which scientists were writing about women's biology. But first she would critique her academic institution.

Arditti started meeting with a group of other women who worked at Harvard, and they made plans to produce what became a seventy-five-page booklet, *How Harvard Rules Women*. Although they wrote and produced the work collectively, Arditti composed the section on Harvard Medical School, describing its pompous architecture and the paintings of big men adorning its halls. The spirit of sexism had long permeated the institution, she said.[28] When a woman applied in 1850, male students argued, "No woman of true delicacy would be willing in the presence of men to listen to the discussions of the subjects that necessarily come under consideration of the student of medicine." The faculty refused her application. Nearly one hundred years later, in 1945, when Ruth Hubbard was contemplating applying, the first women were finally admitted to study medicine at Harvard. But that didn't mean their road was easy.[29]

The nine women who were writing *How Harvard Rules Women* often met at Arditti's apartment on Donnell Street in Cambridge. On warm afternoons, they would sit on the porch, looking out onto the leafy green street, enjoying each other's company while nine-year-old Federico played in the living room. Betsey Useem, a PhD student at the Harvard School of Education, recalled those meetings. She remembered looking up to Arditti, who was older than most of the other women in the group yet had created a comfortable, happy place for dialogue, friendship, and work. Given her international background, she seemed more worldly and sophisticated than some of the others, and they loved the cadence of her Argentinian accent. She also had nuanced opinions so was less doctrinaire than some of the other feminists at the time.

In June 1970, the group printed and released *How Harvard Rules Women*, which they sold for seventy-five cents a copy. They deliberately chose to remain anonymous, preferring to work as an egalitarian collective rather than highlight individual authorship.[30] In the prologue they quoted Harvard's dean of freshmen, F. Skiddy Von Stade, who claimed that the Radcliffe women who were protesting the Vietnam War were "the worst of the bunch." He had respect for the male students because they were at risk; if they were expelled, they would likely be drafted. But if the women students were thrown out, they would "just go off to secretarial school." In response, the pamphlet authors pointed out that a woman with a bachelor's degree from Radcliffe still might struggle to find a secretarial or low-level technical job; and they emphasized that many women, with or without a degree, served as unpaid household labor. Notably, however, the authors did not address how Black women fit into this scenario.[31]

Arditti created a table for the pamphlet showing that Harvard Medical School's class of 1970 was comprised of 111 white men, thirteen white women, one Black man, and no Black women. She pointed out that the white women in the class were generally treated as "a high-risk group of undesirables." Then she contrasted the student body with the employees at the medical school: out of a total of 1,056 staff members, 815 were women. She did not break down the numbers according to race or ethnicity, but she did note that these women were not only secretaries, librarians, and cleaners but also lab technicians who furnished data and conducted the tedious parts of experiments so that men could publish the results.

In her chapter text, Arditti noted that, for many years, the lunch room had traditionally been reserved for students and professors, not for staff or employees. No one questioned this situation until one Black woman secretary, whom Arditti did not name, courageously took the initiative and pushed the school to allow secretaries and technicians access to the lunchroom.[32]

The authors of *How Harvard Rules Women* believed that the liberation of women was not simply a matter of getting more women into leadership positions. It meant breaking down myths, such as the claim that women didn't have the stamina to run the Boston Marathon—a biologically specious conclusion, given the 1966 groundbreaking performance of the runner Roberta

Gibb.[33] It also meant breaking down the way in which class, family structures, and the economic system were oppressing women.

At about the time that *How Harvard Rules Women* was released, Rita Arditti traveled to Cold Spring Harbor to deliver a paper to the annual Symposium on Quantitative Biology. The focus of that year's symposium was the transcription of genetic material. More than three hundred people were present in the hall— nearly 90 percent of them men—as Arditti spoke about the in vitro transcription of lactose genes. James Watson, the molecular biologist who had helped to explain the structure of DNA, was present. Also in the audience was Arditti's young son Federico, who sat in the front row, watching trees and raindrops beyond the window. He looked up at his mother with pride.[34]

Among the few women listening to Arditti's presentation was Barbara McClintock, whom Evelyn Keller had once glimpsed at Cold Spring Harbor. She had been based at the laboratory since 1941—the same year that Hubbard had enrolled at Radcliffe—doing research on maize genetics. Though McClintock had made important contributions in the 1940s and 1950s, her work had not been fully recognized. By the time Arditti was giving her talk, McClintock was generally regarded as an old-fashioned biologist. Unlike Arditti and Watson, she wasn't involved in the new field of molecular biology.

On another rainy afternoon during the symposium, Arditti showed her colleagues two films that she had acquired through Newsreel, an activist filmmakers' organization. The first was *People's Park*, about protesters in Berkeley, California, who in May 1969 had tried to establish a public park on land owned by the University of California and had been brutally repressed by the police. The second was an interview with Bobby Seale about the Black Panther Party. About a hundred scientists watched the films, and many made donations to cover the screening costs and stayed afterward to talk about the antiwar movement and Science for the People.

As one scientist later told Arditti for an article in the magazine *Science for the People*, it felt incredible to talk about such topics at Cold Spring Harbor.[35] Until this point, the facility had rigorously avoided politics. In the previous year's annual report, for instance, Watson had described the Cold Spring Harbor campus as one hundred acres of quietly beautiful shorefront "without the distraction of good or bad films, drug stores carrying paperbacks, or bars with TVs if not topless waitresses." In her later report for *Science for the People*, Arditti pointed out that his male chauvinist quip had put waitresses into the same category as films, books, and television.[36]

Returning to Cambridge, Arditti worked on the editorial board that was readying the August 1970 publication of the first issue of *Science for the People*, which would have a print run of 5,000 copies and cost fifty cents. Emblazoned on its cover was the organization's signature power fist and flask, and Arditti had convinced her colleagues to include the resolution demanding equality for women in science that had been unofficially tabled at the AAAS council session.[37] In December 1969, several hundred AAAS members had signed the resolution, and a member of the organization's Committee on Council Affairs had suggested that the women's caucus of Science for the People send the statement to *Science*, the association's official magazine. Those editors had rejected it via form letter, but now it would be more widely disseminated.[38]

Still, this wasn't the resolution's first foray into print. An earlier SESPA newsletter had published a short synopsis, and even that abridged version had generated ire. A Mr. Khanduri of Washington, D.C., wrote, "Men and women are two complementary aspects of life and it is extreme stupidity to compare the two. They are like day and night; right and left. They have been made for different purposes. Women should stay at home, be mothers, housewives, etc. I can write a big article on this subject. So please withdraw this question of 'Equality of Women in Science.'"[39]

Rita's political awareness was provoked by the Vietnam War. Only later did she become involved in the women's movement. But once these doors were opened, she felt as if the entire house had to be torn down and rebuilt. Her son Federico still remembers this transformation in his mother—from a middle-class scientist who conformed to society's rules to a passionate activist who cared deeply about social justice.[40]

And her work was having an impact. The AAAS decided to hold a panel on "Women in Science" at their next annual meeting, in December 1970.[41] Organizers contacted Ruth Hubbard to ask if she would give a talk about her life as a female scientist. Hubbard accepted the invitation and, as part of her preparation, invited ten other women scientists to her house in Cambridge for a group conversation about their experiences.

The group gathered together on a handwoven Mexican rug in the living room, which Wald and Hubbard had filled with art picked up during their travels, including a large Acoma pot and a West African mask. A print of Picasso's dove of peace hung on the wall. The visitors included Betty Twarog, a neurophysiologist who had discovered serotonin in the mammalian brain while working at the Cleveland Clinic and who had since moved to Cambridge.[42] Also present was Zella Hurwitz Luria, a psychologist at nearby Tufts University. Alice Huang from Harvard Medical School was there, as were Ursula Goodenough, a postgraduate student who worked in the Harvard biology department with Hubbard, and Carolyn Cohen, a biophysicist with a PhD from MIT, who was a cancer researcher at the Children's Cancer Research Foundation (today called the Dana Farber Cancer Institute).[43]

Sitting in lotus pose, Hubbard described the demonstration at the women's luncheon at the AAAS conference and asked her listeners what they thought. As they began to respond, the women started to realize that all of them had been insulted by their male colleagues and their profession. Each one of them had been frustrated by stagnation in their career.

As Hubbard looked around the room, she suddenly realized that this was the first time she had intentionally met with other women scientists rather than incidentally happening to be in the same room with them, often because of whom they were married to. Each of these women was highly accomplished, but all acknowledged that they had marginal jobs—"nonjobs," as Hubbard would later call them.[44] They held titles such as *associate* or *lecturer* and had little job security. Men of their age and with similar levels of accomplishment were in entirely different positions—tenured or on their way to becoming professors.

The conversation that day upset many of the women. But it was also cathartic. "It was the closest thing I had ever had to a religious experience," recalled Ursula Goodenough. Sitting on Hubbard's living-room floor, she felt that every truth she'd once accepted as a given had now been "subjected to a 180-degree spin." The invitation to Hubbard's house became the catalyst for her own emerging feminism.[45]

The women met weekly for six months. They called the discussions *consciousness raising*, an approach that, as Arditti had learned during her first foray into feminism, was becoming popular in women's groups around the country. The experience was entirely new for Hubbard, who had always preferred the company of men. It was the first time she'd been a part of a women's group of any kind.[46]

Another first was confronting the discrimination she was experiencing in academia and science. She thought about the members of the group—their positions as women in science, their relative lack of influence and recognition.

As she later said, "It was pretty mind blowing for all of us . . . and it made us angry."[47]

During the summer of 1970, Hubbard began to reflect in her personal journal about how her life as a scientist had been shaped by society's expectations of her as a woman. She saw herself as a scientist first, then as Wald's wife and her children's mother. Yet society didn't see her that way: "[No one] expect[s] me to have an identity apart from my husband and children, and I do. I feel I'm being drowned in the 'wife of . . .' bit, also probably because once one begins to think about these things, they begin to rankle worse than before." Hubbard's resentment was building. She recognized that, after her marriage, and especially after Wald's Nobel Prize and his MIT speech, her identity and scientific contributions had been subsumed by her husband's work and interests. She was unhappy and knew she had to make a change.[48]

To this point, the women's movement had not had much of an impact on Hubbard. She, along with other women on campus, had often taken an apologetic tone when describing their situations at Harvard. Maybe if they were careful and said the right thing, things would get better. Maybe things weren't all that bad.[49]

When Rita Arditti had first shared *How Harvard Rules Women* with Hubbard, Hubbard had paged through it and then tossed it aside. But now both women were starting to use their growing feminist awareness to look at science in new ways.[50]

By 1969, Evelyn Fox Keller and her husband had moved to California for his sabbatical year, but she was continuing to work in her field. She had recently published a journal article with Lee Segel that excited not only biologists and chemists but also physicists and mathematicians. Keller and Segel were trying to find an explanation for how slime mold amoeba aggregate. If you spread these single-celled organisms onto a plate of agar and give them bacteria to consume, they will busily eat and multiply, splitting in half to create new single-celled organisms. But when the food runs out, the amoebae reorganize themselves in multicellular clusters. Over time these clusters differentiate with a head and a tail so that the organism can move in search of food. How amazing is this, and how does it happen? wondered Keller and Segel. They created a model that offered an explanation. Their article was enthusiastically received, and it continues to be widely cited more than fifty years later.[51]

Despite other scientists' great interest in her research, Keller couldn't find a job in California. Without work or affiliation, she wondered about her future. Each week, Segel, her lab partner and co-author, would phone her with great excitement from New York. He was being invited everywhere to give presentations, and their article was greeted with enthusiasm each time he presented it. Keller couldn't help but feel despondent.[52]

She began to wonder why there were so few women in science in the United States. While her son was at school and her daughter was at afternoon daycare, she spent time at the library gathering data about the history of American women scientists. She had known that their numbers were small, but now she was surprised to learn that these numbers had dwindled even more during the past few decades. She wondered why. And was there a connection between her own feelings and how other women scientists felt about their work?

At first Keller thought the answer might lie in individual psychology. Not once did it occur to her that her disenchantment might be related to the fact that she was not sharing in Segel's kudos. Before her eyes, their joint work was becoming known as his, with Keller seen as his assistant.[53]

During a visit to the East Coast, she attended a Friday-night lecture at the Marine Biological Laboratory in Woods Hole on slime mold aggregation, the topic of her famous paper. The speaker was Aharon Katzir-Katchalsky, a renowned biophysicist, and he spoke specifically about Keller and Segel's work. After his talk, she hurried to the front of the room to introduce herself and suggested that they schedule a time to talk. "I'm so sorry," he said. "I'm only here for a short time and I'm much too busy." But at 8 a.m. the next morning her phone rang. It was Katzir-Katchalsky. "Salva Luria told me you are E. F. Keller. I didn't know!" he said. This insult was not an isolated event. There had been other times when Keller had tried to talk to people who had written or lectured about her work. They'd ignored her. They had not realized that E. F. Keller was a woman.[54]

PART TWO
FEMINIST CRITIQUES
OF SCIENCE,
1970 TO 1990

CHAPTER 4

TURNING A FEMINIST EYE
TO SCIENCE

As Rita Arditti became more engaged in the women's movement, she also began pondering her work situation. For close to five years, she had been working as a research associate in Jon Beckwith's lab at Harvard Medical School. She realized that her career was at a dead end. Once she'd believed that sexism didn't apply to her, that it was a personal matter that affected individual situations. Now she was coming to conclude that she and many of her friends and colleagues were operating within a patriarchal framework. She was beginning to see how the personal, the professional, and the political were entwined. Yet when she suggested to Beckwith that there should be support structures for the people who worked in the lab, he looked at her as if she had lost her mind.[1]

In the early 1970s, Arditti was one of three women serving on the medical school's Commission on the Status of Women, a group that, ironically, was chaired by a man, Albert Coons. Arditti and her friend Colleen Meyer, also on the commission, attended all of the meetings, taking notes and participating actively. After each session, they would discuss the mens' behavior—interrupting the women, not taking them seriously.[2]

Arditti started looking for a job that would allow her to build her experience in teaching and interact with people outside of the lab. She found an opportunity as an assistant professor, teaching first-year biology to non-science majors at Boston University. For the moment, she also continued to work part time in Beckwith's lab. Stepping into the world of teaching was a challenge; and at the end of the first semester, she admitted that her students had been bored because the course had emphasized the memorization of facts and names and had presented scientific work as if it had happened in a vacuum, separate from the social and political conditions of the times, a common approach at most colleges and universities. Without establishing a connection between scientific knowledge and real life, Arditti had struggled to inspire or energize students. She knew she needed to breathe new life into science for these undergrads.[3]

Working with Tom Strunk, a cellular physiologist, Arditti put together a new syllabus for the second semester. They called the class "Biology and Social Issues," and its topics included genetic engineering, cloning, birth control, sterilization, abortion, and questions about the biological basis of behavior. A newly released book, *Our Bodies, Ourselves* (1970), was on the list of recommended readings.

The class was so popular that Strunk and Arditti decided to design another new course for the following year. Called "Objecting to Objectivity," it was one of the first attempts to combine biology and social issues in a science class. Student interest was overwhelming; Arditti and Strunk had to limit registration to 140 participants. During the sessions, Arditti directly told her students that she was starting to question elitism and the position of women in science. She also talked about the class with her colleagues at Science for the People, telling them that it was allowing her to unite her laboratory knowledge with her political beliefs.[4] In a course evaluation, one student wrote, "It was easily the first science course I've had that actually had some practical value." Another said that it had changed his outlook on science completely. He had begun the course with an image of science as Mendel, Curie, and Einstein, but the sessions had made him think instead about science's potential to both solve and create human problems. Many students were grateful for a class that had encouraged enquiry rather than memorization.[5]

Meanwhile, in the fall of 1973, members of the Boston chapter of Science for the People decided that they needed to look more closely at problems of "sexism, racism and elitism" in science. The chapter held two meetings on these topics, but the discussions were dominated by men and focused predominantly on class struggle. Arditti and the other women attendees were dismayed that they hadn't gotten a chance to discuss sexism and other issues of concern among women scientists. They decided it was time to end their isolation from one another and formed their own women's issues subgroup.[6]

Approximately eighteen women, among them Arditti, Freda Salzman from the physics department at the Boston campus of the University of Massachusetts, and Lila Leibowitz from the anthropology department at Northeastern University, formed the core group. This original group soon broke into two. One was a study group that focused on reading and learning and formed an alliance with socialist feminists, and over time its members began to put together specific action plans. The other served as a support group for women, helping them deal with immediate problems of sexism in the workplace and within Science for the People. Each splinter group met separately but came together about once a month.[7]

The women's issues group was comprised predominantly of white middle-class women, and some members were aware that this was a problem. In an article submitted to the September 1974 issue of *Science for the People*, two of them, Carol Axelrod and Ruth Crocker, argued that it was important to consider the point of view of the most oppressed people in any group, whether they be white women, poor women, or Black women. While most white members had not fully embraced the fact that women of color faced different barriers, they were acknowledging the need to broaden their analysis.[8]

Women across the United States were now looking closely at the ways in which history, psychology, and literature portrayed women, but very few had considered science. Although all of the members of the women's issues group understood that they were dealing with the expectations, assumptions, and ignorance of the male establishment, Arditti was one of the first to publish articles in *Science for the People* that took a feminist perspective. "Why are women always being measured against men?" she asked in "Women's Biology in a Man's World," published in the July 1973 issue. "Are men 'normal' and women biological deviants?" She pointed out that, in the late nineteenth century, when American white women were becoming dissatisfied with traditional sex roles and beginning to be involved in education, social movements, and birth control, physicians had predicted "disaster for the species," declaring that "if too much energy went to develop the brain, the ovaries would suffer."[9]

Physicians and scientists had carried this thinking forward into the late twentieth century. In her article, Arditti quoted Bruno Bettelheim, the famous psychologist, as saying, "We must start with the realization that as much as women want to be good scientists and engineers, they want first and foremost to be womanly companions of men and to be mothers." Arditti argued that this ideology had prepared the ground for "natural scientists to propose androcentric (male-centered) theories of human nature and human evolution." She made it clear that she was worried about not only the challenges facing women as scientists but also the androcentric bias in biological theories about women. She was ready to challenge existing scientific assumptions about human nature.[10]

Arditti's article reviewed issues related to hormones, birth control, and the concept of biology as destiny. She closed by asking, "Who runs the entire scientific and technological establishment?" Then she answered her own question: "a small, powerful elite, mainly white males." In her view, the interests of this elite often ran opposite to those of the great majority of people, both men and women.[11]

As Arditti was drafting this article and teaching her new courses, she and twelve other members of the women's issues group had started working on a series of articles for a special issue of *Science for the People* that would focus on how science and ideology had been and still were used to oppress women.[12] Their work on this special issue pushed them to think about scientific racism as well. In a draft introduction, group members wrote about race and IQ, a topic that was garnering public interest. In a 1969 issue of the *Harvard Educational Review*, Arthur Jensen, a professor of educational psychology at the University of California in Berkeley, had claimed that about 80 percent of individual intelligence is due to genetics and that the disparity in IQ scores between Blacks and whites in the United States was linked to genetic differences. Thus, in his view, social support programs such as Head Start were a waste of time and money.[13]

In the explosive debate that followed, Richard Herrnstein, a Harvard psychology professor, publicly supported Jensen's claims. In 1971, he published a controversial article in the *Atlantic Monthly* that drew on Jensen's work and concluded that IQ was largely determined by genetics. Responding to Herrnstein's article, members of Harvard's branch of SDS retorted sarcastically in the *Harvard Crimson* that Black people apparently "must remain poor because they are genetically inferior."[14]

Jensen's and Herrnstein's articles stirred a debate that would continue for years, and Arditti and her colleagues pondered over why the articles were getting so much attention.[15] They concluded that, at root, the articles and their conclusions were racist. The successes of the civil rights movement had made many people receptive to arguments suggesting that there was scientific support for the status quo. Thus, specious scientific articles were being picked up by the popular media and being used to support racist conclusions.

The group noted the same types of prejudices in articles about women, many of which suggested that there were natural reasons for women to center their social and work roles on the home. After discussion, members formally recorded a statement in their minutes. They declared that there was a clear link between the rise of the women's movement and the rise in so-called scientific articles telling women they were "cut out to be passive and stay home cooking." Scientific commentary was being used to explain the superiority of white men.[16]

These kinds of discussions percolated in Arditti's mind, and her frustration with her work situation grew steadily. Male scientists at Harvard were telling her that she couldn't do good work in the lab if she were only there part time, that her work would not be respected. The reality was difficult to accept: as a woman, Arditti was supposed to be responsible for childcare, teaching, and full-time lab research, and the cards were stacked against her. She had been unable to develop her teaching abilities at Harvard so had looked for opportunities elsewhere, but now she was being criticized for her reduced lab presence. It was time to look for a new job.[17]

Nonetheless, Arditti was becoming prominent in other realms. Impressed by her contributions to How Harvard Rules Women, organizers of a symposium about the lives of women in science invited her to chair a panel at the New York Academy of Science. Afterward, a woman in the audience stood up and said, If you get a good husband, you can do science. Another said, If you have money, it's easier. Arditti was irritated by the dearth of real discussion. No one had addressed the issues of divorced or single women with children. No one had talked about how little support women in science were receiving from mentors and colleagues.

Nevertheless, another woman in the audience, a student at the Union Graduate School in Cincinnati, liked what Arditti had had to say and invited her to lunch. A few months later, Arditti received a call from the woman (whose name she did not record), saying that Union needed women faculty and asking if she would be interested in applying. The school's mission was to offer graduate studies to older students who lived at a distance and needed to work at their own pace with tutorial support, a very different approach from traditional graduate education. Arditti applied, interviewed, and was offered a job. What she learned from the experience was that speaking her mind on a panel could pay off in the end. That had not been true at Harvard Medical School and in science. But this situation was different, and she liked it.[18]

In April 1974, after starting her new job (which allowed her to continue to live in Cambridge), Arditti joined three other women to open New Words, one of the first feminist bookstores in the United States. The goal of New Words, first housed in nearby Somerville and later in Inman Square in Cambridge, was to build a community of women who were eager to talk about books and ideas. Arditti was excited about the new direction of her life: teaching biology at Union, working with Science for the People, launching New Words.[19] In the store's early days, she read every single book on the shelves about women and science. What Arditti was absorbing was the beginning of a wave of writing about science, women, and gender. Within a decade, that body of literature would explode exponentially .

In September 1974, the special women's issue of *Science for the People* was released. In her article "Women as Objects," Arditti described a scientific community composed mainly of white middle-class males who had accepted the myth of the neutrality of science. According to her argument, male scientists had been socialized in a professional value system and often ignored the fact that basic cultural assumptions were having a major impact on their work and the interpretation of their findings. She pointed to assumptions about gender and race as examples.[20]

Historians of science, wrote Arditti, had begun to recognize that scientists were working from the model of science that Francis Bacon had proposed in the sixteenth century. Bacon saw nature as the enemy and science as the instrument of its control and domination. All of the European scientific societies, those elite male institutions, had sprung from Bacon's vision of science. "Clearly," she wrote, "there is no place for women in this scheme except as objects of study."[21] Now, four centuries after Bacon, patriarchy in the United States continued to dehumanize women and set them up as objects. As an example, Arditti showed how men in power had historically reduced females to their reproductive organs. This assumption continued to allow the contemporary scientific community to offer supposedly scientific rationalizations for the "secondary status of women."[22]

Arditti pointed out that the medical view of a healthy human body was male and that for a thousand years medical education had portrayed females as less developed than males. She followed this trend through the

Renaissance and into twentieth-century Europe and the United States. Even in the 1970s, normal, healthy aspects of women's biology—menstruation, pregnancy, menopause—were viewed as "almost pathological episodes from which women w[ould] 'recover.'"[23]

Only a few years earlier, Arditti had not imagined that the women's movement had any implications for science. Now she was wondering how a feminist philosophy of science might develop. In the concluding section of her article, titled "Some Thoughts on Feminism and Science," she argued that a changed science could affirm everyone's identity rather than create a biology of superior and inferior. A more feminist approach to science could shift practitioners from a focus on "exploitative, value-free technology" to an emphasis on the needs of human beings rather than profit. Women would no longer be seen as "reproductive units of the species," and a priority would be placed on women's health.[24]

The reaction of many of Arditti's scientist colleagues was disappointing. They said she was "crazy" to consider these matters as relevant to her work in science. The myth of objectivity in science was so strong that most people thought the questions she and the other women were raising were totally absurd. Colleagues asked, If science is the truth, how can you raise these issues? Arditti was objecting to objectivity.

As Rita Arditti was leaving her job at Harvard Medical School, Ruth Hubbard was thinking more and more about her role as a woman in academia and science. In April 1971, she read the ninety-six-page report of the Committee on the Status of Women in the Faculty of Arts and Sciences at Harvard and its twenty-four detailed recommendations. The first stated that Harvard needed to increase the number of women faculty. But by May 1972, nothing had changed. Outraged, Hubbard and a colleague in her department drew up a document. Signed by seventeen faculty members, it demanded an end to "the present paucity of women and minorities among the Faculty." At the time, many Harvard personnel saw this as a radical statement, and they severely criticized the signatories.[25] But Hubbard stood her ground. Not only were there very few women professors on campus, there were absolutely no African American women on the faculty. An article in the *Harvard Crimson* noted, "The status of women within the University has hovered near the bottom of Harvard's list of priorities for several hundreds of years. And despite appearances, this past year has proved no exception."[26]

Like Arditti, Hubbard was realizing that she wanted to add teaching to her work. Initially, however, she was insecure about her abilities. She began by taking on one section of Wald's introductory biology course. Again, she felt as if she were riding on his coattails. But the course went well, so she accepted a lecturer role in another class, "Biology and Social Issues," teaching a section on the politics of health care and another on the Jensen and Herrnstein controversy about race and IQ. This led to a three-year stint teaching a house seminar—that is, a course outside of the biology department—called "The Politics of Health Care."

Hubbard now had two titles: research associate and lecturer. In contrast, her male contemporaries were either on the ladder to professorships or had already received them. This was beginning to seriously bother her. Once she had been naïve enough to think that she was being saved from attending dull administrative meetings. But now she was in her mid-forties. Her children were growing up, and she had no security except through Wald. She wondered if she could get tenure in the biology department. When she started talking about the matter to her colleagues, some said she should have been considered long ago.[27]

Hubbard spoke to the dean and asked to be considered for tenure. Though he didn't dismiss the suggestion, nothing moved forward, and Hubbard wondered what she should try next. For a short while, she taught science at a school in Roxbury, a predominantly Black area of Boston. She also spent time doing pregnancy counseling, offering support to women who needed to travel to New York for an abortion. (The procedure was still illegal in Massachusetts.)[28]

Meanwhile, as Hubbard's feminist awareness grew, she, like Arditti, slowly started to question scientific knowledge. She could see that the women's movement was affecting literature, history, and psychology. Women in these disciplines were asking questions about how their fields had been shaped by men and examining the roles that women had played in them. Hubbard started asking similar questions about biology. She approached the issue as a scientist, examining data and looking for ways to test her ideas, and decided that the best place to start was evolutionary theory.

Hubbard hadn't reread Darwin's *On the Origin of Species* in decades and decided to take another look, all the while keeping a specific question in mind: does this theory show male bias? She quickly realized that she couldn't have picked a better book. It was evident that Victorian England had had a significant impact on Darwin's scientific thinking. In his telling, all males compete for access to fertile females, and all females sit on the sidelines as if they are at a dance, each trying to choose the best male to improve her situation. It was blatantly clear that Darwin was bringing his own cultural beliefs into his scientific analysis. As Caroline Kennard had done ninety years earlier, Hubbard questioned Darwin's conclusions and began developing new ways of thinking about science.[29] She backed away from her laboratory work and allowed herself to reflect on broader questions of the philosophy of science. She asked, What is it that we scientists really find out about nature? If scientists were more representative of society in terms of gender, race, and class, would science look different?

Having grown up in a socialist home, Hubbard had always identified with the left. Yet she had needed the impetus of two major social upheavals—the Vietnam War and the women's movement—before she was able to recognize that scientists' work was greatly affected by their social and personal backgrounds. The realization hit her like a thunderbolt: science had been made almost exclusively by white, highly educated, economically privileged men. It was one of the most liberating insights she ever had.[30]

It wasn't the only change she went through at this time. In a letter to a friend, she discussed growing up in a homophobic society and said that the women's movement had educated her on issues facing the gay and lesbian community. Among other influences, she recalled that, in 1972, one of her closest female friends had shared with her that she was having a love affair with a woman. Along with her feminism, Hubbard became an advocate for queer rights and began to develop friendships with lesbian students and serve as a mentor. Their perspectives and challenges had a significant impact on her political awareness.[31]

Hubbard also had an awakening about racism. She later recalled the moment, in the late 1960s, when she'd first been able to discuss race and racism with an African American friend. In later remarks at a college commencement, she spoke of being amazed that she'd been well into her forties before reaching this point.[32] Yet while she credited the civil rights movement with broadening her perspective, she also realized that she had spent much of the decade focused on her work and her young children, not thinking deeply about public affairs. Only when her brother Sascha became a lawyer for the Black Panther Party did she begin to understand the scope of the situation. Sascha's connection with Huey Newton and Bobby Seale had an influence on Hubbard's politics, giving her the impetus to speak out against racism in the 1970s, when the Boston area was engulfed in controversy over busing and school integration.[33]

During the uproar about the racist implications of Jensen's article on race and IQ, Hubbard paid attention. In 1973 and 1974, Harvard's Committee against Racism held a series of forums to publicize and oppose academic theories that promoted the idea that Black people naturally had lower IQs than white people.[34] Although Hubbard hadn't been involved in the women's special issue of *Science for the People*, she now read through it and agreed that the timing of this debate was significant. Like members of the women's issues group, she, too, had observed that research on sex differences was entering the media mainstream as a response to second-wave feminism.[35] As movements promoting racial equality, women's equality, and gay rights gained strength, scientists were increasingly focused on questions about racial differences and sex differences. Their answers to these questions tended to be conservative because the scientists themselves were predominantly white and male and had a strong social interest in maintaining the status quo.[36]

Ruth had a realization: just because something is biological does not mean it cannot be changed. Our biology is changing all the time; it constantly responds to how we live. She thought back to elementary school, when she felt awkward and uncoordinated, struggling to climb a ladder and walk across a balance beam. After a woman gymnastics teacher had helped her feel more confident in her body, she had exceled in team sports. Many factors influence an organism and lead to changes in biology. Biology is not destiny.[37]

As Hubbard was thinking about science in new ways, things were also shifting on the political front. In March 1972, President Richard Nixon signed the Equal Opportunity Act. Finally, women academics who had been suffering under discriminatory hiring practices, settling for unequal pay, and overlooked for tenure would have some recourse in the courts. With the looming possibility of legal investigations, several universities, including Harvard, decided to appoint a few senior women to tenure-track positions. The easiest way to do this would be to transform long-marginalized lecturers into full professors.[38]

"We should have appointed you ten years ago," said Dean Henry Rosovsky as Hubbard's new position was being finalized. He was not altogether happy about the move and directly said that he was concerned that Hubbard was past her prime, a remark he probably would not have made to a newly hired man. She took it to mean "Know your place: we are doing you a favor."[39] At Harvard, professors receive a gift when they get tenure—a black wooden armchair embossed with the university's "Veritas" crest: the tenure chair. After twenty-five years of service, they receive a similar chair. Hubbard was awarded both chairs in the same year. Instead of celebrating them, however, she thought of them as just another place to sit.

A reporter later asked her, "Do you think that being a woman had something to do with you being appointed?" Hubbard responded, "I don't know whether my being a woman had something to do with my getting tenure in 1973, but I'm sure it had something to do with my not even having been considered before." She was the first woman ever to achieve tenure in biology at Harvard, and she was about to turn fifty.[40]

Hubbard had traveled to Vienna once before, briefly, as an adult. But after receiving tenure, she made a longer visit. She bought a map and a guidebook, walked the streets, and rode the tram. She sat in sidewalk cafés and marveled at how the women were dressed—high heels and flouncy skirts, powder and lipstick—so unlike the pants and Birkenstocks that she was wearing. They looked as if they were still in the 1950s, she thought. Even so, as she wandered through the city, she discovered that she felt much more European than American. She thought back to 1938, the year her family had fled from Austria, and wrote in her journal, "[My] rejection of Vienna crippled my soul. Now that I have let it back in, I can try to be whole again." For a moment, she wondered if she should leave Wald, leave Harvard, and resettle in Europe. Not long before this trip, her first husband, Frank Hubbard, had suddenly died in his mid-fifties. She felt sad and uncertain. Where did she belong? Where was she going? Should she continue to do lab research and publish scientific journal articles? Or should she head in a new direction?[41]

Hubbard wrote a poem to her teenage children, asking if they were feeling strong in themselves. She thought back to her own teen years, when she was "running to grow up" and trying desperately to forget who she was, even before she had a clear idea of her identity.[42] Yet despite her uncertainty, achieving tenure had given her concrete security. Rather than focusing only on research, data, and publishing, she now had the opportunity to step back and think. She decided to develop a new seminar, one she would call "Biology and Women's Issues"—a vague name because she was still considering exactly what she wanted to explore. But whatever the details would become, she knew she wanted to investigate scientific explanations for biology from a feminist perspective.[43]

In 1975, Hubbard taught her first iteration of the course to ten students, all of them women. Several had already done some reading around the topic, and a number were involved in the women's movement. One, Mary Sue Henifin, later recalled a few of the readings that Hubbard assigned that year: Frantz Fanon's *The Wretched of the Earth*, Paulo Freire's *Pedagogy of the Oppressed*, and Mary Daly's *Beyond God the Father*. They read about the history of science. They read Darwin.[44] Alongside her students, Hubbard explored nineteenth-century explanations of women's biology. They discovered that scientists such as Darwin had found so-called "natural" explanations for social assumptions about women, that doctors had described menstruation and the uterus as controls over women and had declared that reproductive functions were the central purpose of their lives. The students learned that Edward Clarke, teaching at Harvard in the 1870s, had told women not to pursue higher education but to conserve their energy for bearing children.

In the course's second year, three men showed up at the first class, along with ten or eleven women. As Hubbard recalled, the disconnect in knowledge was immediately evident: all of the women students had done a lot of reading on their own, but the men had not. As the students talked about their expectations for the course, she realized that some of the women would drop out if the men were to stay; the course would be too simplified for them. Hubbard spoke to the male students, offering them an independent-study option so that they could get through the readings at their own pace. However, none of them took up the offer; and when a fourth male student heard about the situation, he cried discrimination and took his complaint to the university and the media.[45]

After reviewing the case, a faculty council decided that the facts were too vague: they couldn't tell if the men had withdrawn because they'd been excluded as male or if they hadn't had the necessary background for an advanced course on women's biology. Nonetheless, a long article in the *Harvard Independent* suggested that Hubbard's course had violated Title IX, one of the Department of Health, Education, and Welfare's 1972 education amendments, because it had discriminated based on sex. A counter-article in the *Harvard Crimson* scoffed, "The whole question was overblown to the point of absurdity." Ursula Goodenough, who had become chair of the Faculty Standing Committee on the Status of Women, agreed: "It's ridiculous that this should be the first Title IX case at Harvard, particularly when there's been so much discrimination from the other side."[46] Hubbard decided to take a different tack when she next offered the course. She was prepared to accept everyone who registered and would discuss the matter with the group. But this time, no men showed up.

In its fourth year of existence, "Biology and Women's Issues" became a standard departmental course rather than an elective seminar. It continued to gain in popularity, even attracting a small number of men, and Hubbard offered it every year until her retirement in 1990. At first, she treated it as a traditional class in which she was the leader and in control of most of the presentations, though students did present on topics that they were researching and write a final paper. Yet she and the class members soon realized that the student papers were extremely valuable on their own. No one wanted to file them away in a box. So, in the late 1970s, inspired by her reading of Darwin and by the quality of the student research, Hubbard started working on an essay titled "Have Only Men Evolved?"

At this time, there was next to no published research that focused on a feminist analysis of science. Hubbard decided to pull together a collection of essays that would include some of the papers from the course, and she invited two seniors—Mary Sue Henifin and Barbara Fried—to be her co-editors. The result was *Women Look at Biology Looking at Women*, first published in 1979.[47]

Meanwhile, Hubbard's daughter, Deborah Wald, was growing out of childhood and working to define her role in the world as a young woman. Hubbard was attentive to this, and Debbie felt that her own life experiences were enriching her mother's work. The two were undergoing parallel struggles. As Hubbard was trying to get tenure at Harvard, Debbie was attempting to get a girls' sports program going at her school. Though the battles were different, to mother and daughter they felt like part of the same struggle.[48]

Deborah later recalled bringing home a pamphlet from the school nurse and reading it with her mother.. The booklet explained how girls entering puberty could "handle the odors and embarrassment of menstruation" without compromising their femininity, and both mother and daughter were outraged. Hubbard encouraged Debbie to do research that would give her the ability to push back against such stereotypes, and she bought her a copy of *Our Bodies, Ourselves*.

Debbie and Elijah Wald were close to the age of Rita Arditti's son Federico, and for a time they all went to the same school. Their mothers were already loosely acquainted as colleagues, but gradually they got to know each other better through their children. Arditti felt exhilaration and relief when she learned that Hubbard was beginning to ask feminist questions about science. There were so few women scientists, and even fewer who were open to this perspective. She said to herself, Another woman in science who sees these issues! She was even happier when, after reading the special women's issue of *Science for the People*, Hubbard attended a few meetings with the women's issues group.[49]

In 1974, Rita Arditti was thirty-nine. That May, as she was standing in front of the mirror, she noticed a dimple on her right breast. When she touched it, she felt a hard spot. She kept checking for a few days, hoping it would go away, but when it didn't, she saw her doctor. He told her to consult a surgeon. Without knowing whether or not she had cancer, the surgeon recommended a modified radical mastectomy. Arditti stalled and insisted on a biopsy.[50]

A few days after the procedure, the surgeon phoned to say that, indeed, the spot was a carcinoma. He wanted Arditti to go into the hospital for surgery the next day. Again, she stalled. Her son Federico would graduate from elementary school that day, and there was no way she was going to miss it. She postponed the surgery till after the event.

In the mid-1970s, women had little access to information about breast cancer. People didn't even say the word *cancer* out loud. Post-mastectomy, Arditti was shocked at the sight of her chest. She felt like "a big scar." But the ordeal wasn't over. As soon as the incision had healed, she started radiation treatment at MIT. Technicians strapped her into a chair and fixed her head against a pole. The chair swung back and forth in front of the radiation source as she stared at the travel posters of Paris, Amsterdam, and Switzerland that someone had hung on the walls. She felt like an astronaut in a space capsule.[51]

At the end of the radiation treatment, Arditti's doctor was optimistic. He told her that she was probably "cured," but she was scared, and she had to begin the long process of learning to live with the fear that her cancer would return. She was also worried about her new position at Union Graduate School. Should she tell her employer about the cancer? What if she lost her job? Fortunately, she had strong support from her feminist friends at New Words and in the women's issues group of Science for the People, and she had the satisfaction of seeing her article "Women as Objects: Science and Sexual Politics" appear in the special issue of *Science for the People*.[52]

The Hubbard-Wald family was shocked by Arditti's diagnosis, and Elijah was especially concerned about her son. Federico was thirteen years old, an only child, dependent on a single mother. If his mother died, he would have to move to Europe to live with his father, which he desperately did not want to do.[53] He was terrified, and Arditti was frank with him. She told him she was going to handle her own medical care and treatment and that she didn't know what would happen next.[54] One day, when she was out of the house, Federico got down on his knees next to his mother's bed and prayed. He made a deal with himself and with God: "If my mother lives, I promise, I will always live nearby."[55]

Amazingly, Rita lived for another thirty-five years. Despite a diagnosis of stage 4 cancer, she drew on her knowledge as a cellular biologist to manage her own treatment. But in 1974, there was a great deal of uncertainty ahead.

CHAPTER 5

SHE PROVED THEM WRONG

In 1974, Evelynn Hammonds was studying at Spelman, a historically Black women's college in Atlanta. While Ruth Hubbard and Rita Arditti had both experienced antisemitism during their childhoods, and both had come to recognize that sexism was affecting their lives as scientists, they were exposed to racism in society only through the civil rights movement, the Black Panther Party, and the controversy around race and IQ. Having grown up in segregated Atlanta, Evelynn Hammonds's world was different.

One day that spring, she walked across the green quadrangle of Morehouse College, a historically Black men's school, affiliated with Spelman where she was taking her physics classes. She entered the red-brick building of the physics department, pushed open the heavy wooden door into her classroom, and sat down among a group of students gathered at the front of the room. As she took her seat, she noticed that two Morehouse professors were sitting at the back.

The students were gathering to elect officials for the Morehouse and Spelman chapter of the national Society of Physics Students. By a show of hands, they elected a male student as president, a female student as vice-president, and another young woman as treasurer. The next open position was secretary, and a woman student put her name forward. Suddenly one of the young men raised an objection. "You're telling me there's going to be three women in the leadership?" he shouted. "We can't have that."

Hammonds flinched. She wondered what his problem was. The world was changing. As far as she knew, women had every right to be in this room and every right to lead. Nevertheless, she could feel the tension rising.

Another man, one of Hammonds's teaching assistants, leaned forward in his chair and started to complain. "They're taking over," he said. She recalled that he had recently handed back an assignment to her with a handwritten note saying, "You can't be a black woman and a physicist." Hammonds was already mad at him about that, and now she was seething.

The election for secretary deadlocked. One of the men pushed for a roll-call vote—that is, asking each person, one by one, for a verbal vote—and the group agreed. But when one of the men voted for the woman candidate, another shouted, "What is wrong with you?" Someone else piped up, "You can't do this." Meanwhile, the two professors at the back of the room stayed quiet. Why don't they say something? thought Hammonds. It occurred to her that there were no women faculty in the physics department. The roll call continued, and no one else—including the professors—voted for the woman candidate. In the end, the male candidate was elected as secretary. What was going on?[1]

"Evelynn, I really want you to understand," said Carl Spight, the department chair at Morehouse, when she went to talk to him after the election. "It's not that I don't think a woman would be good. But you can see that I couldn't vote for a woman."[2]

"Why not?" she demanded and then walked out, not waiting for his answer.

"You can't be a woman *and* a scientist," her boyfriend explained as they walked across campus.

Hammonds was stunned. "What is your problem?" she snapped back.

"You don't understand," he said. "You're going to have to choose."

"What are you talking about?" she asked. She stayed calm, but her jaw tightened and she quickened her step.[3]

It had never occurred to Hammonds that she would have to choose. She knew there weren't very many women scientists, but so what? Times had changed. It was the 1970s: she was wearing Converse sneakers and an afro to class, and women could do whatever they wanted in the world.

It had not dawned on Hammonds that many people saw science as a gendered profession, as inherently masculine. Her parents had been candid with her: they warned her that people would want to stop her from doing what she wanted to do because she was Black. But it had never occurred to her that there were some things that she couldn't do because she was a woman.

Hammonds headed to the library and asked for a book about Black women scientists. She was told there wasn't one. She looked through every book and encyclopedia. To her astonishment, there was nothing. And there wasn't much about white women scientists either. I'm in trouble, she thought. This is not going to be easy.[4]

That year, the physicist Shirley Jackson visited Spelman and spoke to the Society of Physics Students. In 1973, Jackson had completed her PhD in physics at MIT—the first Black woman to earn a doctorate at the institute in any subject.[5] The students watched in awe as she filled the blackboard with complicated equations from her thesis. They were deeply moved by her clarity and confidence when she described her work. Before this moment, Hammonds had never met, or even heard of, an African American woman physicist.[6]

Jackson recounted two major events in her youth that had contributed to her interest in science. One was the Supreme Court's 1954 *Brown v. Board of Education* decision, which opened new educational opportunities for African Americans. The other was the Soviet Union's 1957 launch of *Sputnik*, the first artificial satellite to orbit Earth, which prompted American concerns about falling behind in science, math, and the space race.

Jackson's visit created an uproar in the department. Was it really possible to be a woman and a scientist at the same time?[7] Hammonds went to talk to Spight about her dilemma. Jackson was clearly proof that a woman could be a scientist, but what about the election and her dead-end search in the library? "Don't worry about it," Spight said. "It's not really an issue." But to Hammonds, it was a very big issue—and it was personal too. Her drive was making her unpopular.

"A lot of people don't like taking class with you," a classmate told her. "Because you're so competitive. You're really difficult."

"Just because you come over on a Wednesday night and I say I have to work," Hammond retorted. "Then I'm difficult?" But her heart was racing, and her anger and frustration simmered just below the surface.[8]

One problem was that she didn't have a way to talk about sexism in the classroom, so she started reading about the women's movement and feminism—Shulamith Firestone's *The Dialectic of Sex*, Kate Millett's *Sexual*

Politics, Germaine Greer's *The Female Eunuch*, all published in 1970. It helped to discover that she wasn't the only frustrated woman in the world, but all of these writers were white women. In the mid-1970s, there were few available critiques by Black women. Still, her reading steeled her determination. If people were going to have a hard time dealing with her as a woman scientist, that was "just too bad."[9]

During the 1940s, both of Hammonds's parents had attended Morris Brown College, a small historically Black Methodist school in Atlanta and the first educational institution in the state to be owned by African Americans. Her mother, Evelyn Baker Hammonds, worked as an elementary school teacher. Her father, William Emmett Hammonds Jr., had wanted to be an engineer, but in Jim Crow Georgia there was no place for a Black person to study engineering. Instead, he studied chemistry and mathematics and became a postal worker.[10]

When Evelynn Hammonds was nine, her father gave her a chemistry set. The two of them loved doing experiments together, arranging test tubes and vials of dry chemicals on the kitchen table. When she was ten, he gave her a microscope, and she spent hours staring at the leaves and insects she'd brought in from the yard. When she and her younger sister got watches as presents, she took her sister's watch apart to see how it worked. Emmett Hammonds hadn't had the chance to study engineering himself, but perhaps, he thought, his daughter would. The 1954 Supreme Court ruling had opened educational opportunities for her that hadn't been there for him.[11]

By the time Evelynn was entering ninth grade, Atlanta schools had desegregated, so she took the bus to what had formerly been an all-white school. She was the only Black student in her math class, but she was eager and regularly raised her hand. One day, the teacher looked over at her and said, "I see we have a question. Let's do a roll call and see who else has questions." The teacher went through each person's name on the register, one by one, skipping Hammonds's. Finally, at the end of the roll call, the teacher asked, "Now what is *your* question, Evelynn?" She felt the humiliation and the racism and kept quiet. She knew the math teacher would take every opportunity to give her an A- when she deserved an A.[12]

In 1971, Hammonds enrolled in Spelman College's dual-degree program. Her goal was to spend two and a half years studying physics and two and a

half years at Georgia Tech studying engineering, ending up with two bachelor degrees. The environment at Spelman was enriching, and her professors
supported her drive to achieve. But at parties, when she told young men that
she was an engineering major, they lost interest and looked for other girls to
talk to. At one point, Hammonds started saying she was an English major,
just to avoid debate. A Georgia Tech professor wrote on one of her papers,
"You write so neatly. Why not think of being a secretary?"[13]

As Hammonds later recalled, Black southern women at Spelman were
expected to achieve, but within certain boundaries. The expectation was to
work in the community and better the lives of Black people but also to have
children, raise a family, and achieve all this with "a particular flair." Although
Hammonds's mother, sister and many of her friends wanted her to be successful, they weren't comfortable about challenging the institutions around
women's roles. She said, "I knew that I was doing something very different
within my Black community, to want to be a scientist."[14]

Yet Hammonds continued to feel frustrated in her science classes. At
Georgia Tech, her lab partner would set up the circuit they needed for
their experiment. Then he would turn to Hammonds, hand her a pen, and
say, "Evelynn, you can take down the data." One day, she got to class early
so that she could set up the experiment herself. She knew she needed to
learn how everything worked. So, following the instructions in the class
notes, she connected the jumble of wires, the voltmeter, and the other components. But when her partner arrived, he didn't even glance at her work.
Friendly and chatting, he simply dismantled it and put it together again
himself. "Evelynn," he said, "I'll run the experiment. You can take down the
data."[15]

Hammonds's friends argued that her lab partner was being "gracious and
gentlemanly." But Hammonds didn't agree and reported him to her teacher.
The response: "He's being helpful to you. You seem like a good team." Her
sister told her she was always complaining about what some guy had done
and why it wasn't acceptable.[16] "You must be a dyke," said a young man at
Georgia Tech. Hammonds had never heard the word before and didn't know
what he was talking about. She assumed it must be a rude word he was using
to get her to shut up.[17]

In the summers of 1974 and 1975, Hammonds worked at Bell Laboratories
in a research program for minorities and women. She was assigned to a scientist and was very serious about her work. Hammonds loved being in the lab,
and her devotion paid off: she co-authored a published paper based on her

summer research. The encouragement she received at Bell Labs compensated for the difficult environment at Georgia Tech.

But one day, when Hammonds arrived at the library, she found that the librarian had set aside a new report for her: "The Double Bind: The Price of Being a Minority Woman in Science." Immediately, she sat down at a table and read the entire eighty-page document. This was what she had been looking for. The report described a remarkable gathering that had taken place in December 1975, when thirty women of color, including Shirley Jackson, had convened outside of Washington, D.C., for a conference hosted by AAAS. At the event, the women had spoken about how efforts to encourage women in science were generally focused on white women, whereas efforts to support people of color were typically focused on men. The report had been co-authored by Shirley Malcom, Paula Quick Hall, and Janet Welsh Brown, all three of whom would go on to make important contributions. Malcom, who had earned a PhD in ecology from Pennsylvania State University, was a new employee at AAAS. She would spend the next fifty years working to implement the report's recommendations.[18]

Inspired, Hammonds wrote a letter to Malcom. "I want to be a physicist. Where should I go to graduate school?" she asked.

"I think you should apply to MIT," Malcom responded, and reminded Hammonds that Jackson had studied there.[19]

With Malcom's encouragement, Hammonds applied to MIT for graduate study in physics. Not only was she accepted, but she was also awarded a scholarship from Xerox; and in the fall of 1976, she made the big move from Atlanta to Cambridge.

The main entrance to MIT is imposing, with columns, a neoclassical dome, and limestone steps. As Hammonds walked into the foyer and peered up into the five-story-high dome, she felt excited. Inside the building she strolled down what was known as the Infinite Corridor, passing classrooms, offices, labs, and noticeboards. Outside were tall trees and vast green lawns, where students were reading, playing Frisbee, and chatting. She looked out across the Charles River that divides Cambridge from Boston and had a clear view of the city skyline. Hammonds had arrived in a region with more than fifty colleges and universities and hundreds of thousands of students, and she was eager to join them. Here, she thought, it would be okay for her to be smart. There were more women here, too—women who had read about feminism and become involved in the women's movement. She hoped that her growing identity as both a feminist and a scientist could flourish at MIT.[20]

Hammonds's first setback took place during her opening week of classes, when her adviser suggested that she had gaps in her training so should start by taking a freshman physics course. Hammonds declined; in her view, this was not a sensible suggestion. She knew another student from Georgia Tech who was also attending MIT. He, too, had some educational gaps to fill but had been given more reasonable advice: to take a junior- or senior-level undergraduate course in the areas he needed to compensate for. Hammonds was angry. To start all over again with freshman physics made no sense. Not only that, the suggestion was racist. Her anger stayed with her, and she didn't have anyone to guide her through it. She'd only been in Cambridge for a week, and already she felt embattled.[21]

She tried to keep the example of Shirley Jackson in mind. In the late 1960s and early 1970s, Jackson had been the first Black woman to survive the MIT physics department. It had helped that she'd been able to work with the only Black physics professor, James Earl Young. But her field was different from Hammonds's: Jackson had studied high-energy theoretical physics, whereas Hammonds was studying experimental condensed-matter physics. Even by the mid-1970s, MIT had very few Black graduate students on campus, and Hammonds was one of only two Black women in her department. Officially, she had an adviser, but she didn't feel that anyone was acting as a mentor.[22]

As her feminist awareness grew, Hammonds struggled more and more to deal with men's dismissive remarks about women scientists, faculty members, and teaching assistants. One afternoon, for instance, a woman scientist gave a talk in Hammonds's class. None of the men sitting close to Hammonds had anything to say about the content of the lecture, but they had plenty to say about how the woman was dressed. Hammonds vowed to herself that, whenever she gave her first talk, she would wear pants, not a dress, and make sure that everyone in the audience was paying attention to her words.

Hammonds realized she would have to be better than the men in her class if she were to get anywhere in a scientific career. As a Black woman, she figured she would have to be at least six times better. This was tremendous pressure, and she was incensed at the blatant misogyny of the men who worked alongside her.[23] In one of the teaching labs, the male students had tacked up a few photos on the walls, mostly of women in bikinis, including the famous poster of the actress Farah Fawcett in a red one-piece bathing suit, head thrown back, tawny hair tangled. After considering these images, Hammonds decided to add to the gallery. She taped up a colorful poster of a woman in a head wrap who was carrying a baby on her back and a rifle in her hands. Printed along one side were words from a poem by Yosano Akiko: "The mountain-moving day is coming. I say so, yet others doubt. . . . All sleeping women, now will awake and move."[24]

The lab instructor, who was also her adviser, Bob Birgeneau, took one look at the new addition and said, "No more posters allowed up in the lab." Clearly, Hammonds's poster had prompted the command, but Birgeneau knew better than to single her out. "Okay, fine," she told him. "I'll take my poster down as long as everybody else takes theirs down." They did.[25]

One day, Hammonds found a copy of an early edition of *Our Bodies, Ourselves* lying in a classroom. Randomly, she opened it to chapter 2, "The Anatomy and Physiology of Sexuality and Reproduction," and was stunned by what she saw—drawings of the vulva and the clitoris. In her experience such information about women's bodies was not talked about, except in whispers. As she paged through the chapters—

"In Amerika They Call Us Dykes," "Taking Care of Ourselves," "Rape," "Venereal Disease," "Birth Control," "Abortion"—she thought about the women she'd grown up with in Atlanta. Her mother and her aunts did not discuss such things even in private. The students at Spelman did not talk about abortion, rape, or domestic violence.[26]

Our Bodies, Ourselves was a radical document. While Hammonds was grateful that it was opening up conversations about women's health, she noticed that it included very little information about Black women. She began thinking hard about the absence of Black women in other areas linked to science, women's issues, and public life.[27]

During Hammonds's first year at MIT, she spent a lot of time with the other Black woman in her class, but eventually that friend found the environment too difficult and left.[28] Now Hammonds became the only Black woman in the department, though one of her professors briefly had a Black woman secretary. One day this professor walked into Hammonds's office at nine in the morning. "You have to retype this," he said and threw some papers on her desk. She had no idea what he was talking about. A few minutes later, he returned, looking embarrassed. He told her he had mistaken her for the secretary. Yet throughout the year he had repeatedly seen her in his classes and in the hallways.[29]

As a physics graduate student, Hammonds felt invisible. She was forced to prove herself at every turn—not because she wanted to be one of the boys but because she wanted to be taken seriously. She thought back to the men who said "you can't be a woman and a scientist" and wanted to prove them wrong. Why aren't there more women here? she wondered. Why aren't there more Black people?

Hammonds's dog-eared copy of "The Double Bind" was always in her bag, a reminder that she could achieve what she wanted to achieve.[30] Still, nearly every day she had to endure comments that were overtly racist or sexist or both. Every day she questioned why she was still here. The comments were infuriating, but anger was draining her energy and disrupting her ability to focus on her studies. She had to figure out a survival strategy.[31]

So she concentrated on what she did well: her lab work. Hammonds was learning about the structure of various materials by analyzing the scattering of X-ray beams, a technique that Rosalind Franklin had used in the 1950s when she was studying the structure of DNA. The process was long and painstaking, but it allowed Hammonds to do cutting-edge work on her master's thesis topic: the X-ray scattering of xenon (a substance used in flash-bulbs and other specialist lamps) as adsorbed onto graphite.

This research had practical implications for materials science, and Hammonds received significant recognition for her work.[32] Nonetheless, she felt defeated. She'd been locked in battle since the moment she'd arrived at MIT. She decided against pursuing a PhD. The departmental culture and the lack of support were too heavy a burden. She would leave academia and find a job. However, she knew she was a scientist, and she had earned her master's degree in physics.[33]

Years later, an interviewer asked Hammonds if sexism had been the over-riding problem at MIT or if racism had affected her more. She answered:

> They are not separate, because they are not separate in me. I am always Black and female. I can't say, Well, that was a sexist remark, without wondering, would he have made the same sexist remark to a white woman? That takes a lot of energy, to be constantly trying to figure out which one it is. I don't do that. I just take it as somebody has issues with me and who I am in the world. Me being Black, female and wanting to do science. That's it.[34]

In 1974, as Hammonds was first discovering what she would need to endure in order to be a Black woman scientist, Evelyn Fox Keller was invited to give a series of lectures at the University of Maryland on her work in mathematical biology and physics. The invitation was an honor and she wanted to accept it, but she had a caveat. Keller had just finished teaching her first women's studies course. During those sessions, her students and colleagues had encouraged her to talk about what it had been like to become a woman scientist. Before this time, she had only shared her experiences in private. But now she saw the story as more than personal, as something that had broader significance. The personal was becoming political. [35]

Keller felt that there were reasons to publicly share her own experiences and those of other women in science. She was troubled by the way in which apologists for the status quo often claimed that intrinsic differences between the sexes were an explanation for disparities in advancement. Her research had convinced her that there was no evidence of biologically determined differences in intelligence or cognitive styles between males and females, no matter how hard people were working to find such proof.[36]

So, as she prepared for the lectures at the University of Maryland, Keller made a courageous decision. She admitted that she no longer felt comfortable talking about her work in science as if it had been accomplished in a vacuum. In her final talk in the series, she would review the many obstacles against women scientists and discuss possible solutions for removing them. In her view, one of the most problematic barriers was the common assumption that science and scientific thought were intrinsically masculine. Where had this ideology come from, and why was it lingering in a discipline that claimed to be objective and neutral? What consequences did it have on the actual doing of science?[37]

Keller was also realizing that this social assumption could have a major influence on children. For example, one day her daughter Sarah's nursery school teacher asked the children, "What does your father do?" When Keller objected to this one-sided questioning, the teacher asked the children what their mothers did. Sarah responded, "My mother cooks, she sews, she cleans, and she takes care of us."

"But, Sarah, isn't your mother a scientist?" asked the teacher.

"Oh, yes," said Sarah.[38]

In her talk, Keller noted that it was not surprising that even very young children were liable to acquire the belief that certain activities are male and others female. The view of science as male was so tenacious that it was resistant to change, continuing into adolescence and beyond. She herself had been told that a girl who thinks clearly thinks "like a man."[39]

Keller pointed out a myriad of obstacles that blocked women from careers in science. One of the largest was the way in which institutions and society reacted to childbearing. Women often had children soon after obtaining their

PhDs; but "after having been out of a field for a few years, they usually find it next to impossible to return to their field except in the lowest level positions. I need hardly enumerate the additional practical difficulties involved in combining a scientific (or any other) career with the raising of children." Keller argued that "equality of the sexes in the work and professional force is not a realistic possibility until the sex roles in the family are radically redefined." Yet, at the same time, "our society does not have a place for unmarried women. They are among the most isolated, ostracized groups of our culture."[40] In her talk, Keller described what had happened to her in California in 1969, when Lee Segel was raking in accolades for their joint work and she was struggling to find a job. She told her listeners that her own situation was only one example "of a rather ubiquitous tendency, to give more public recognition to a man's accomplishments than to a woman's accomplishments."[41]

Keller's points were important and valid, yet in the mid-1970s they were not seen as a proper topic for academic or scientific discussion. Still, even though she was aware that she had violated professional protocol, she felt profoundly liberated. The moment had been such an epiphany that she decided to publish her lecture, and it appeared as "Women in Science: An Analysis of a Social Problem" in the October 1974 issue of *Harvard Magazine*.

This marked the beginning of Keller's work as a feminist critic of science. She had made two shifts in her thinking. She had begun to pay attention to *gender ideology*—that is, beliefs about male and female nature—and had also begun to see that these beliefs could affect science. Once, such conclusions would have been unthinkable for her, but the women's movement and feminist theory were offering Keller new ways of looking at her discipline.[42] Years later Keller said that she'd had to dig deep to find the courage to give a lecture questioning the neutrality of science, then dig deep again to write her biographical essay "The Anomaly of a Woman in Physics," which described her experiences in the Harvard physics department in the late 1950s. But doing so had given her the momentum to formulate the question that would occupy her for the next two decades: how exactly does the ideological association between masculinity and science affect the field and shape scientific work?

Keller's essay "The Anomaly of a Woman in Physics" was published in 1977 in the anthology *Working It Out*, which included several pieces in which women spoke about their professional lives. Soon afterward, Keller got into an elevator packed with male scientists and one female colleague. The woman said loudly, "Oh, Evelyn, how does it feel to go naked in public?"[43] This was no surprise. Even by the late 1970s, few women in the scientific community would risk sharing their personal experiences. Not only was such talk seen as unprofessional, but many feared it would jeopardize their professional image and their objectivity. Yet Keller embraced her new political consciousness as a source of strength. She'd concluded the essay with the hope that the political awareness rising from the women's movement would support young women to challenge the pervasive idea that "certain kinds of thought are the prerogative of men."[44]

Keller's essay had an impact on many women scientists, including Evelynn Hammonds. Reading it, she was reminded of her first day at MIT, when she'd received the harsh advice to take the freshman physics course because she'd graduated from a historically Black college. The two women's circumstances and histories were different—Keller had arrived in Cambridge in the 1950s dealing with entrenched misogyny; Hammonds had faced both the indignity of racism and the indignity of sexism. Neither had been welcomed into physics, but both persisted as scientists.

In 1970, Margaret Rossiter, who had grown up in Malden, Massachusetts, was one of only a few women at Yale University who was studying for a PhD in the history of science. Rossiter was reading about the women's movement, and she wondered why she was hearing so little about women in the field. During her classes, none of her professors ever mentioned a woman scientist.

Every Friday afternoon, the professors and students in her graduate program, mostly men, would get together to drink a few beers. Although she wasn't particularly fond of their pipe smoking or style of humor, Rossiter tried to attend so she wouldn't be dismissed as uninterested. One afternoon, during a lull in the conversation, she put down her beer and asked, "Were there ever any women scientists?"

The answer from her professors was unequivocal: "No. There have never been any."

Another student asked, "What about Marie Curie?" He was referring to the famous Polish-French scientist who had won the Nobel Prize twice, once in physics (1903), once in chemistry (1911). But the professors rejected even Curie, claiming that she'd merely supported her husband's genius and scientific achievements.[45]

Clearly, the history of women in science was not an acceptable topic among this group of men. Rossiter didn't argue with them, but the conversation set her thinking. And in 1972, after she completed her dissertation on the history of agricultural science, she went to work. As she paged through the directory *American Men of Science*, she was astonished to discover many entries for women. Women scientists had been present all along, though even experienced historians of science had never noticed them. She later wrote, "I felt like a modern Alice who had fallen down a rabbit hole into a wonderland of the history of science that was familiar in some respects but distorted and alien in many others."[46]

Still in her twenties, Rossiter began what would become her life's work: bringing to light the history of women in science in the United States. She wanted to understand how their stories connected with the rest of recorded history and why they had been forgotten. But she got very little support from other historians. "There is nothing to study": this is what several scholars told her. They were convinced that American women scientists were a recent and still very isolated phenomenon.

Rossiter quickly found proof that this wasn't true. *American Men of Science* included the biographies of more than five hundred women scientists who had been at work in the early 1900s. As she read these entries, she came to realize that the discrimination they had faced then was not so different from what women were facing in the 1970s. For instance, although the majority of women scientists in the early twentieth century were more highly educated than their male counterparts, they had fewer job opportunities, faced higher unemployment, and held positions of lower status. Yet, as Rossiter also discovered, many of these women were also working hard to overcome such discrimination.[47]

Rossiter tried to publish some of her initial research, but top journals such as *Science* and *Scientific American* were not interested. Eventually, she placed an article in the May 1974 issue of the less prestigious *American Scientist*. Not surprisingly, neither scientists nor historians paid much attention. But Rossiter persevered, and in 1982 she published her full findings in her landmark book *Women Scientists in America*. Its release took Ruth Hubbard and Rita Arditti by surprise. They had been unaware that, throughout the 1970s, Rossiter had been gathering historical information about women in science. During that period, they and others in Science for the People had been more occupied with current events—including speaking out against a new field of study: sociobiology.

CHAPTER 6

QUESTIONING THE SCIENCE OF GENDER

The headline was "Sociobiology: Updating Darwin on Behavior," and it was on the front page of the *New York Times* on May 28, 1975. The article discussed the book *Sociobiology: The New Synthesis*, which was garnering major attention for its author, E. O. Wilson, whose photo, bespectacled, in jacket and tie, was also prominently featured on the front page. The headline and photo caught Ruth Hubbard's attention, and she began to read the article with growing unease. She'd long known Wilson as a colleague; he too, was in Harvard's biology department. Throughout the 1950s and 1960s, while she'd been in the lab studying the biochemistry of vision, Wilson had been studying the social life of ants. His 1971 book, *The Insect Societies*, had described his research on their classification and behavior. Even then, he'd been speculating that the same principles might explain the behavior of other animals. This new book had taken that leap, and Hubbard was deeply skeptical.

The article described a new field called sociobiology, which Wilson defined as "the systematic study of the biological basis of all social behavior."[1] The "revolutionary implication," according to the reporter, was that much human behavior, including aggression and altruism, could be "as much a product of evolution as is the structure of the hand or the size of the brain." The writer mentioned that the new book was already stimulating excitement amongst the biologists who had seen advance copies.[2]

Hubbard had been rereading Darwin. Based on what data, she wondered, had Wilson come to this conclusion about humans? When she got her hands on a copy of the seven-hundred-page book, she saw that he had indeed extended his analysis in *The Insect Societies* to include many kinds of animals. Using the tools of population genetics such as kinship selection and inclusive fitness theory, he was attempting to explain why natural selection could explain certain behavior. He explored dominance systems, roles and castes, sexual selection, and parental care as aspects of behavior that were shaped by evolutionary biology.

Most of the book focused on animal behavior, both vertebrate and invertebrate. In more than five hundred pages, he discussed ants, termites, starlings, dolphins, rhesus monkeys, and baboons. However, the book's subtitle, *The New Synthesis*, was a reference to his suggestion that the social sciences could be seen as branches of biology because biology could explain human behavior. With this notion in mind, he chose to open the book with "The Morality of the Gene," a nod to the human social sciences, and close with "Man: From Sociobiology to Sociology," arguing that human cultural variation was not free of genetic influence.[3]

Hubbard agreed that biology and environment—nature and nurture—both play a role in shaping human behavior. But she was concerned about some of the questions that Wilson was asking as well as their implications. "A key question of human biology," he wrote, "is whether there exists a genetic predisposition to enter certain classes and to play certain roles." Recently, Hubbard had been thinking about how people's assumptions can shape the questions they ask in science. She had also been thinking about the malleable nature of human biology. So she was not convinced by Wilson's argument about human behavior, especially when the entire database for his research was based on the behavior of ants and other animals. But she had no idea how heated the sociobiology debate would become.

Not long after reading the article, Hubbard had a conversation with Richard Lewontin, a population geneticist who was also a colleague in her department. He told her that Jonathan Beckwith from the Harvard Medical School (in whose lab Rita Arditti had worked for five years) was reaching out to members of Science for the People, hoping they would be willing to discuss the growing public interest in sociobiology. Hubbard said she would like to join them.

Toward the end of July, the meeting took place at Beckwith's house in Cambridge—outside on the porch so that everyone could enjoy the breeze.[4]

Hubbard spoke up right away. She noted that a growing number of zoologists and evolutionary biologists were becoming sociobiology enthusiasts and said that the accumulating praise for *both* the book and the concept was cause for concern. The group agreed that this was a problem. They wanted to take action, to critique the "screen of approval" that was going up around the idea of sociobiology.[5]

Many had already been involved in countering the Jensen-Herrnstein arguments about IQ, genetics, and race and had spoken out against scientific rationales for male aggression. Their discussion that evening focused on their worry that, like these other theories, sociobiology would be used to spread the idea that biology dictates social outcomes—that society's problems could be blamed on the people who are victims of social and economic inequality, that the problems were in their genes. The group was especially concerned that sociobiology would be used as scientific support for racist social policies. Hubbard also worried that it would prop up sexist beliefs and social policies, although her colleagues didn't see women's issues as top priorities.

In early August 1975, the *New York Review of Books* published C. H. Waddington's positive review of *Sociobiology*. In it, Waddington suggested that findings on biology's relation to behavior would be the biggest development in the field since molecular biology in the 1950s—an attempt to synthesize the entirety of current knowledge in biology and the social sciences. The group that had met on Beckwith's porch was not happy about the review. They decided to craft a response that would push their views about biological determinism and racism into the public discourse. Stephen Jay Gould, a paleontologist and historian of science at Harvard, wrote a first draft that became the basis for group discussion and revisions. The final letter, signed by sixteen members of the group, including Ruth Hubbard, was published in the *New York Review of Books* on November 13, 1975. That same month, under the aegis of the Genetic Engineering Group of Science for the People, they published "Sociobiology: The Skewed Synthesis" in *Science for the People*. Not long afterward they changed their name to the Sociobiology Study Group.

In their letter to the *New York Review of Books*, group members noted that scientists had long been proclaiming that natural selection determines the most important characteristics of human behavior. Versions of this theory stretched back to Herbert Spencer (who had coined the phrase "survival of the fittest") and forward to Konrad Lorenz, Robert Ardrey, and now E. O. Wilson. "Each time these ideas have resurfaced the claim has been made that they were based on new scientific information," but what these biological determinist theories have actually done is consistently provide a genetic justification for the status quo. In the early 1900s, they had been the foundation for sterilization laws and immigration laws in the United States and eugenics policies that led to "the establishment of gas chambers in Nazi Germany." Wilson, the writers said, was joining a long line of biological determinists whose work had exonerated governments of any responsibility for social problems because these problems supposedly existed as a result of innate biological factors.[6]

As group members discussed the various drafts of their public statements, Hubbard pointed out that Wilson had turned to evolutionary biology and ancient human origins to explain sex differences, gender roles, and family structure. For instance, he'd written:

> The building block of nearly all human societies is the nuclear family. During the day the women and children remain in the residential area while the men forage for game or its symbolic equivalent in the form of barter and money. . . . What we can conclude with some degree of confidence is that primitive men lived in small territorial groups, within which males were dominant over females.[7]

Hubbard wondered how Wilson thought he knew this as fact. She was underwhelmed by his conclusion and determined to challenge his patriarchal and sexist views.[8]

She and the other group members had another reason to worry. Jon Beckwith's wife, Barbara Beckwith, who was a high school teacher, had discovered that her school was using a biology curriculum that had been developed by the sociobiologists Robert Trivers and Irven DeVore, Harvard colleagues of E. O. Wilson. The text posited questions such as "Why don't females compete?" and "Why aren't males choosy?" This curriculum, which promoted biological reasons for human behavior and sex roles, was being used in more than a hundred school systems in twenty-six states.[9] In other words, sociobiological theories were now infiltrating public school education.

Signatories to the *New York Review of Books* letter included both Beckwiths, Hubbard, Gould, and Lewontin—a strong showing of Harvard-linked colleagues. Wilson's rebuttal, also published in the review, called it an openly partisan attack. In his view, the signatories were mistaken in identifying an underlying political message in his book. He rejected the idea that sociobiology was the latest attempt to reinvigorate theories that had promoted sterilization laws and eugenics and called such claims "ugly, irresponsible and totally false." He saw the letter as intimidation, an attempt to "[diminish] the spirit of free inquiry and discussion."[10]

The argument that any critique of his work was an attack on academic freedom was a defense that Wilson and his supporters would use for many years to come. He set himself up as a victim, declaring that he'd had no idea that his colleagues were preparing such a letter until three days before it was published. He wrote of being saddened by this attack from formerly friendly colleagues.

Hubbard, for one, was not moved by these emotional appeals. For years, she'd felt alienated from many of her department colleagues, finding them patriarchal and elitist. When it came to important political questions, institutional loyalty to Harvard was not her priority. In her opinion, the stakes were too high to worry about maintaining smooth collegial relations. This became clear in July 1976, when Hubbard and Wald took a public stand against the university, speaking to the Cambridge city council to protest the potential social and ethical dangers of recombinant DNA research in the city, including the possibility that organisms could be created that would be harmful to humans and the environment. They called for public oversight and regulation, but the Harvard establishment saw their action as a betrayal.[11]

By his own admission, Wilson had paid little attention to the activist left in the Boston area. He said he'd never heard of Science for the People and wrote that the forceful objections made by the Sociobiology Study Group had taken him "by surprise."[12] Nonetheless, interest in the study group was growing. Within the first few months of 1976, it expanded to thirty-five members, many of them scientists, psychologists, philosophers, and students. Among them was Rita Arditti, even though she was juggling a new job, the bookstore, and her health concerns. Other new members included the chemist Marian Lowe, the anthropologist Lila Leibowitz, and the physicist Freda Salzman, all of whom had worked with Arditti on the special women's issue published in *Science for the People* in 1974.

Lowe had been the first and was still the only woman on the chemistry faculty at Boston University. There were few such jobs open to women. Earlier, when she'd applied for a position at a university in the South, she was told directly, "We won't hire you because we don't hire women."[13]

Leibowitz was an associate professor at Northeastern University. Her doctoral thesis had argued that the human family "is an artifact of culture," and she was well known for her article "Desmond Morris Is Wrong about Breasts, Buttocks, and Body Hair," a retort to Morris's 1969 *The Naked Ape*, which claimed that the female form was a product of male preference. As an anthropologist, she couldn't accept his view that social and cultural differences have no impact on the relationship between men and women, and she was very concerned about sociobiology.[14]

In 1965, Salzman and her husband George Salzman had been hired as founding members of the University of Massachusetts Boston's physics department. But in 1967, the university's anti-nepotism policy started making things difficult for her. These policies, which had been implemented at most major American universities in the 1950s, were designed to protect universities against having to hire incompetent family members. In reality, however, they hurt many women's chances of academic employment. For five years, the Salzmans fought for Freda's position, and in May 1972 she was finally rehired.[15]

During one study group meeting, Freda Salzman read aloud from a letter that a graduate student in anthropology had written to E. O. Wilson. The student had just purchased *Sociobiology*, and she was writing to tell him that she couldn't get past one particular paragraph, in the closing chapter, which refers to changes in the female estrus cycle. According to Wilson, while females of some primate species experience slight bleeding, only human females have "a heavy sloughing of the wall of the 'disappointed womb' with consequent heavy bleeding."[16] The student wrote:

> You must have had good reasons for putting the words "disappointed womb" in quotation marks. Whether or not you personally sympathize with that sulky organ I have no way of knowing. What I *do* know is that you have drawn my attention to the plight of billions upon billions of "disappointed sperm" dying like so many teensy-weensy beached whales on the sands of a bedsheet. Honestly, I couldn't sleep a wink all night for thinking about them.[17]

Hubbard, Salzman, and others thought the letter was hilarious, but they also agreed that it was clearly pointing out what they already knew: that Wilson's views were biased. While his books did not explicitly advocate for patriarchy, he did present heterosexuality, the nuclear family, male aggression, and female maternity as natural and maintained that these human behaviors resulted from natural selection and evolution. The other women in the study group, several of whom were starting to develop feminist critiques of evolutionary theory, recognized that sociobiology was trouble for women.

Although growing numbers of women were taking public issue with Wilson's claims, the sociobiology debate quickly became framed as a battle between men. At first, media coverage focused on debates between Wilson and Richard Lewontin, who was often referred to as the leader of the Sociobiology Study Group. After the British author Richard Dawkins published his own pro-sociobiology book, *The Selfish Gene*, in 1976, the media reframed the debate, centering it around Dawkins and Stephen Jay Gould.

In scholarly circles, the concerns of women scientists such as Hubbard, Arditti, Leibowitz, Salzman, and Lowe received limited attention. This skewed coverage became apparent in 2000, when Ullica Segerstrale published *Defenders of the Truth*, which many sociobiologists, including Wilson and Dawkins, saw as the definitive book about the debate. Segerstrale had been a sociology student at Harvard in the mid-1970s and had had a front-row seat to events as they'd happened. Yet her description of the debate focused almost exclusively on the men's perspective. Though her book mentions Ruth Hubbard in passing, it never alludes to any of the many other women who were taking part in the discussion. In a review of *Defenders of the Truth*, the biologist Anne Fausto-Sterling pointed out that the book was a document of "he said—she said," except that there was barely any "she said" at all.[18]

While Wilson spoke openly about sex differences, he said very little about race. Though he'd been influenced by Jensen's and Herrnstein's writings on IQ and race, he didn't want to light that fuse in his book. Instead, he wrote that IQ is only one subset of the components of intelligence and suggested that the genes that contribute to creativity, entrepreneurship, drive, and mental stamina are likely scattered across many chromosomes. "Even so, the influence of genetic factors toward the assumption of certain *broad* roles cannot be discounted."[19]

Publicly, Wilson distanced himself from controversial books of the 1960s such as Morris's *The Naked Ape* and Robert Ardrey's *The Territorial Imperative*. He explicitly framed his arguments as more modern, more scientifically based, though he had been corresponding with Ardrey for years and took great interest in his books.[20] Yet Wilson did not hold back from pointed comments about women in *On Human Nature* (1978), which focused solely on the sociobiology of human behavior, won a Pulitzer Prize, and was widely read. In it, he declared that evolutionary factors were behind the fact that not a single country currently had a woman head of state. Even if women were to have identical educational opportunities and equal access to all professions, "men are likely to maintain disproportionate representation in political life, business and science."[21]

Wilson died in 2021. Perhaps he would have been surprised to learn that, as of September 2022, the world had thirty female heads of state.

Thanks to Arditti, Salzman, and others, Science for the People had developed an awareness of the problems facing women and science. But, like other parts of society, the organization had not overcome sexism. The attention it gave to issues facing women in science was inconsistent, and the women in the Sociobiology Study Group were constantly dealing with internal problems related to gender. They were frustrated, for instance, because the group did not consistently employ gender analysis while debating issues. Eventually, the women began meeting separately as a subgroup.[22] Here, they were able to have long discussions about sociobiology that were healthy, open, friendly, and exciting. Members felt free to explore ideas and learn together. But when they returned to the full group of men and women, the atmosphere was less relaxed and often unpleasant. The men regularly put down the women or flippantly or impatiently rejected their comments.[23]

On one occasion, several women, including Barbara Beckwith, accused the men in the group of sexism and elitism. Lewontin was disturbed and said that this had not been his intent.[24] Soon thereafter he received a phone call from a science writer at the *New York Times* asking about the latest developments in the group. Sensitive to the gender issues that the women had just raised, he told the journalist to talk to another member.

"What do you want me to do?" asked the journalist. "Pick out someone at random?"

"Precisely," Lewontin replied. The journalist hung up.

In Lewontin's view, this anecdote illustrated the informal leadership structure at Science for the People and within the Sociobiology Study Group: no officers, no central committee, no constitution, no bylaws, no structure. Nevertheless, as the women knew, this informal power structure systematically heightened the voices of the men.[25]

The group's difficult gender dynamics continued for several years. According to the minutes, participants had one particularly long, involved, and heated discussion about sex roles and sexism in sociobiology. The women at the meeting, including Salzman and Lowe, had argued that nearly every biological determinist behavior that Wilson had presented involved sex differences.

They pointed specifically to his closing chapter, in which he argued that certain traits of early humans, such as "prolonged maternal care" and males who were "specialized for hunting," had led to later social evolution. Wilson then drew the conclusion that males have different strategies for reproduction that lead to traits such as aggression, territoriality, and dominance. Men in the Sociobiology Study Group, including Steve Chorover and Stephen Jay Gould, argued that "sexism is only one aspect of many" that leads to this conclusion. They believed that the group needed to focus on a broad range of aspects of biological determinism, not only sexism, in order to reach a broader audience. The women, however, were frustrated that the men did not consider women a broad-enough audience in and of themselves.[26]

In December 1976, Science for the People hosted a showing of *Sociobiology: What Comes Naturally*, a film made by the Canadian Broadcasting Corporation. Viewers, many of them members of the Sociobiology Study Group, watched as Irven DeVore, Wilson's colleague at Harvard, spoke directly into the camera: "If you look at primates, the female, after she is three or four years of age, is essentially just a baby-producing machine." DeVore continued: "The same has been true of humans, mothers, throughout history." The scene shifted to footage of women college students walking with or sitting next to and looking up at men. One leaned her head on a man's shoulder. DeVore's voice returned: "You don't have to be a scientist to notice that, among humans, men are much more interested in status and politics than women are. Wherever one looks, throughout the vertebrates, all the animals, the primates, . . . one finds males competing for status with each other."[27] The scene shifted to show men boxing and playing football. Hubbard, Salzman, and other women in the audience were deeply troubled. Clearly, a sexist point of view was not only informing the entire field of sociobiology but also leaking into school curricula and the popular media.[28]

By 1977, Science for the People was undergoing a high turnover in women's membership. Many young women had left, frustrated because the organization was not specifically incorporating a feminist perspective or fully addressing gender dynamics. In an interview years later, Arditti described the situation as risky and a "constant struggle" and recalled that she always felt "on the defensive." She said that the men in Science for the People knew that they had to give some attention to issues affecting women, "but it had to be limited to what they wanted and it had to be something that they liked."[29]

The topic came up for discussion at the national conference in Voluntown,

Connecticut, in April 1977, after women called for a separate lunchtime meeting to focus on the issue. In his report on that meeting, Glenn Wargo, a member from Boston, wrote, "Uh oh, are we heading for Marx Brothers vs. Lenin Sisters?" Under the heading "Later: Storm Hits," he described women members as they spoke of feeling alienated, of the lack of public support for their ideas. Wargo wrote, "Substantial agreement . . . as they are nodding. Or are they falling asleep?" His sarcastic tone suggested that he wasn't taking the discussion seriously. But another member, Scott Schneider of Michigan, had a more thoughtful response, agreeing that "the necessity for such a discussion became painfully obvious while listening to some of the comments."[30] The women members of the Sociobiology Study Group continued to deal with the burdens of sexism. The group was organizing forums and panel discussions and producing articles, statements, and brochures on the implications of sociobiology. However, only the women were writing specifically about sociobiology and sex roles; no men took part in preparing those materials, even though the issues were front and center in the mainstream media. In the *New York Times Magazine*, Wilson wrote:

> No solid evidence exists as to when the division of labor appeared in man's ancestors or how resistant to change it might be during the continuing revolution for women's rights. My own guess is that the genetic bias is intense enough to cause a substantial division of labor even in the most free and most egalitarian of future societies.[31]

It was clear to the women in the study group that the scientific establishment and the media picked up on sociobiology's support for traditional sex roles, and the men in the group were not concerned. Second-wave feminism and the gains it had made were under attack.

Still, the women persevered. In March 1977, Freda Salzman and Marian Lowe organized a public forum at Boston University titled "Are Sex Roles Biologically Determined?" One workshop looked specifically at sex-difference research, asking, "Why is this research being done? Why is there such a resurgence of interest in biological modes of explaining differences among social groups, including sex differences? What are the social implications of this research?"[32] Salzman published an article by the same title in the July–August 1977 issue of *Science for the People*. In it, she argued that the reemergence of interest in sex-difference research was part of a general resurgence of biological determinist theories. In her view, these theories were trying to show that inequalities based on class, race, and sex were due to genetic differences rather than societal influences. Yet she also pointed out that any genetic basis for sex differences says nothing about how resistant these traits are to changes in the environment.[33]

As debates about sexism in sociobiology heated up, the women members of the study group were continuing to deal with sexism in their scientific careers. Salzman's had come to a full stop for five years during her fight against the anti-nepotism policy. Then, after she was finally granted tenure, she was diagnosed with breast cancer, and she died in the spring of 1981.

Meanwhile, Leibowitz had applied for a full professorship at Northeastern and had been turned down. In a conversation with Hubbard, she attributed the rejection to the fact that she didn't have any references from prominent academics in sociobiological anthropology, only letters from biologists, anthropologists, sociologists, and women's studies scholars. Scholars who embraced sociobiology had gained influence in the academy as well as in public opinion and in the media.[34]

In August 1977, the newsmagazine *Time* ran a cover story titled "Why You Do What You Do: Sociobiology: A New Theory of Behavior." The accompanying image showed a young white couple draped with marionette strings. The man, dressed in a suit and tie, stood smiling with his arms around the woman's waist. The woman's arms were around the man's shoulders, and she stared blankly ahead. The article focused mostly on the work of the Harvard sociobiologist Robert Trivers but credited Wilson for bringing the new science to public attention. It also mentioned a recent conference on sociobiology attended by academics from across the country and suggested that the discipline's ranks were growing. Notably, it had been Trivers, along with DeVore, the voices of the Canadian film on sociobiology, who had co-authored the high school biology curriculum that had so alarmed Barbara Beckwith.[35]

But soon Ruth Hubbard would have a response.

CHAPTER 7

HAVE ONLY MEN EVOLVED?

Published in 1979, the anthology *Women Look at Biology Looking at Women: A Collection of Feminist Critiques* may have been the first book-length treatment of feminist critiques of science. Ruth Hubbard and her students had spent several years putting together this groundbreaking work. In the introduction, dated September 1977, Hubbard asked several questions about the choices that scientists had been making over time. She noted, for instance, that researchers had paid more attention to the social structure of baboons than to that of chimpanzees or gibbons:

> Could this be because it has been easy to stereotype baboon social behavior as hierarchical, with relatively rigid sex roles? Could it be because chimpanzees have very fluid relationships with one another that are difficult to stereotype by sex? ... Is it an accident that among billions of insect species, those whose social behavior easily conforms to rigid roles are the ones that have caught the imaginations of naturalists from the 19th century onward, using "scientific" language related to slaves and queens, workers and soldiers?[1]

Turning to the subject of humans, she asked, "Is it an accident that scientists have been primarily interested in exploring contraceptive techniques that tamper with the *female* reproductive system?"[2]

The book's essays explored and critiqued the history of research on sex differences in hormones, genes, and the brain. Hubbard pointed out:

Most self-fulfilling theories are devised without intent to defraud, and when they are debunked, they at worst damage the reputations of their authors. When such theories become effective tools for oppression, however, they are social dangers. For example, if some scientists who believe (wish?) that women's mental lives are controlled by the physical demands of their reproductive systems (or that blacks are intellectually inferior) proceed to "prove" these hypotheses by devising the necessary test, asking the right questions, finding appropriate subjects, and then come to the obvious conclusions, sexism (or racism) becomes part of the scientific dogma.[3]

Hubbard and her co-editors argued that, for two centuries in the United States, men had based political decisions on so-called "laws of nature." For example, while men agreed that human society would benefit if women could vote, own property, and pursue an education, they emphasized that this was not nature's plan, given women's reproductive role and their lesser intellect. The message "we reproduce; therefore, we are" had remained largely intact since Darwin. Quoting from Virginia Woolf's *A Room of One's Own*, the co-editors reminded readers of John Langdon Davies's warning: "When children cease to be altogether desirable, women cease to be altogether necessary."[4]

"Science is made by people who live at a specific time in a specific place and whose thought patterns reflect the truths that are accepted by the wider society." So began "Have Only Men Evolved?," the essay on Darwin that Hubbard had started working on in the mid-1970s and that would become one of her foundational texts and a key chapter in the book. Even though Darwin was often portrayed as a rational loner swimming against a religious stream, she argued that many aspects of his theory of evolution dovetailed with the social and political ideas, beliefs, and morals of Victorian Britain. In Hubbard's view, these same notions still dominated contemporary biological thinking about sex differences and sex roles.[5]

Hubbard did not accuse scientists of delusion or dishonesty. Rather, she pointed out that they, like anyone else, find it difficult to see the social biases that are built into their own way of life. She believed that "there is no such thing as objective, value-free science." On the contrary, the science of a particular era is part of its politics, economics, and sociology.

Many scientists and historians had already explored the social origins of Darwin's theory of natural selection, but Hubbard took this exploration a step further. She argued that it made sense that the theory of evolution had arisen in western rather than eastern science. While "a belief that all living forms are related and that there also are deep connections between the living and the non-living has existed through much of recorded history." the Judaeo-Christian belief system set "man (and I mean the male of the species) apart from the rest of nature." Referring to Carl Linnaeus, Hubbard asked, "Did species exist before they were invented by scientists with their predilection for classification and naming? And did the new science, by concentrating on differences which could be used to tell things apart, devalue the similarities that tie them together?"[6]

Growing fossil evidence, she said, made the biblical view of "the special creation of each species untenable. And the question of how living forms merged into one another pressed for an answer."[7] Non-western belief systems "that accepted connectedness and relatedness as givens did not need to confront this question with the same urgency." They already understood that everything was linked, so evidence of evolutionary change did not create a contradiction. Only scientists in the Christian West were flummoxed and needed an explanation.[8]

Even in the nineteenth century, sharp-eyed observers were noting that Darwin had been influenced by his own environment. Hubbard quoted Karl Marx as an example. "It is remarkable," he wrote, "how Darwin recognizes among beasts and plants his English society with its division of labour, competition, opening up of new markets, 'inventions,' and the Malthusian 'struggle for existence.'" Hubbard marveled at a similar conclusion by Friedrich Engels: "The same theories, Engels wrote, are transferred back again from organic nature into history, and now it is claimed that their validity as eternal laws of human society has been proved." In short, "Darwin consciously borrowed from social theorists such as Malthus and Spencer some of the basic concepts of evolutionary theory. Spencer and others promptly used Darwinism to reinforce these very social theories and, in the process, bestowed upon them the force of natural law."[9] The circle was complete.

In her essay, Hubbard turned to the concept of sexual selection. Here again she found evidence that Darwin's scientific theory had been influenced by Victorian social mores. In On the Origin of Species (1859), he described sexual selection as an important aspect of evolution, "a struggle of individuals of one sex, generally males, for the possession of the other sex."[10] He described male bees and ants as "fighting for a particular female who sits by, an apparently unconcerned beholder of the struggle, and then retires with the conqueror."[11] Hubbard drew attention to Darwin's nonobjective language, which "cast [animals] into roles from a Victorian script."[12] Some readers, she said, might defend Darwin's descriptions, and "no one can claim to have solved the important methodological question of how to disembarrass oneself of one's anthropocentric and cultural biases when observing animal behavior, [but] surely one must begin by trying."[13]

In *The Descent of Man* (1871), Darwin began his chapter "Principles of Sexual Selection" by describing an active male pursuing a dormant female: "The male possesses certain organs of sense or locomotion of which the female is quite destitute, or has them more highly developed, in order that he may readily find or reach her." Hubbard was astounded. "Wherever you look among [Darwin's] animals," she said, "eagerly promiscuous males are pursuing females, who peer from behind languidly drooping eyelids to discern the strongest and handsomest. Does it not sound like the wish fulfilment dream of a proper Victorian gentleman?"[14] Hubbard was stunned to read: "Man is more courageous, pugnacious and energetic than woman, and has more inventive genius." Darwin writes:"

> "The chief distinction in the intellectual powers of the two sexes is shown by man's attaining to a higher eminence, in whatever he takes up, than can women—whether requiring deep thought, reason, or imagination or merely the use of the senses and hands. If two lists were made of the most eminent men and women in poetry, painting, sculpture, music (inclusive both of composition and performance), history, science and philosophy, with half a dozen names under each subject, the two lists would not bear comparison. We may also infer . . . that if men are capable of a decided pre-eminence over women in many subjects, the average of mental power in man must be above that of woman."[15]

Darwin argued that this preeminence resulted from men's need to defend "their females" from enemies, to hunt, to create weapons, all of which required "higher mental faculties," characteristics that have been selected for in males over time. "Thus," he declared, "man has ultimately become superior to woman."[16] In Darwin's view, it was good that daughters inherited qualities from both fathers and mothers: "otherwise it is probable that man would have become as superior in mental endowment to women as the peacock is in ornamental plumage to the peahen."[17]

Hubbard was incensed by such assumptions. "Although the ethnocentric bias of Darwinism is widely acknowledged," she wrote, "its blatant sexism—or more correctly, androcentrism (male-centeredness)—is rarely mentioned, presumably because it has not been noticed by Darwin scholars, who have mostly been men." She mentioned two women scholars who *had* taken up the cudgels: Antoinette Brown Blackwell, who wrote the first published feminist critique of Darwin, *The Sexes throughout Nature* (1875); and Eliza Burt Gamble, who critiqued Darwin's male bias during his lifetime in *The Evolution of Woman: An Inquiry into the Dogma of Her Inferiority to Man* (1894). However, both books were barely acknowledged. Thus, Hubbard said, it was important to expose Darwin's androcentrism, "not only for historical reasons, but because it remains an integral and unquestioned part of contemporary biological theories."[18]

Hubbard's essay discussed contemporary views of sexual selection and androcentric descriptions of sex roles among various animals and plants. Observers find it "curious," she said, when the sexes cannot be distinguished clearly. This may be why naturalists tend to focus on "animals whose behavior resembles those human social traits that they would like to interpret as biologically determined and hence out of our control." She quoted Wolfgang Wickler's *The Sexual Code: The Social Behavior of Animals and Men* (1973), which describes the reproductive process of algae and notes that "the mark of male behavior is that the cell actively crawls or swims over to the other; the female cell remains passive."[19] Again, the assumptions were circular: "One starts with the Victorian stereotype of the active male and the passive female, then looks at animals, algae, bacteria, people, and calls all passive behavior feminine, active or goal-oriented behavior masculine. And it works! The Victorian stereotype is biologically determined; even algae behave that way."[20]

Turning to E. O. Wilson's *Sociobiology*, Hubbard noted his emphasis on differences in parental investment as related to eggs and sperm. Wilson wrote:

> One gamete, the egg, is relatively very large and sessile; the other, the sperm, is small and motile. . . . The egg possesses the yolk required to launch the embryo into an advanced state of development. Because it represents a considerable energetic investment on the part of the mother the embryo is often sequestered and protected and sometimes its care is extended into the postnatal period. This is the reason why parental care is normally provided by the female.[21]

Yet, as Hubbard argued, this description fits only some animal species and is inaccurate even for humans. At a time when many women around the globe were both working outside the home and raising children, Wilson's description was reinforcing the patriarchal model of the household. "Clearly," she said, "androcentric biology is busy as ever trying to provide biological 'reasons' for a particular set of human social arrangements."[22]

Darwin introduced his theory of sexual selection in *The Descent of Man*, just as first-wave feminism had begun to coalesce. As *Sociobiology* and related works gained traction in the 1970s, his theory received renewed attention. This, according to Hubbard, was likely related to the rebirth of the women's movement. "When women threaten to enter as equals into the world of affairs, androcentric scientists rally to point out that our *natural* place is in the home."[23] She mentioned Ruth Herschberger's *Adam's Rib* (1948), which made fun of myths about sex differences, and wryly noted that, thirty years after its release, the book was still pertinent. Sexual stereotypes were holding their own in the literature on human evolution though it was a field with few fossil specimens and plenty of room for "investigator bias." Notions of *man the hunter* and *man the toolmaker* reflected the gender of most of the researchers, neatly excluding women from these stories. But as Hubbard argued:

> It makes sense that the gatherers would have known how to hunt the animals they came across; that the hunters gathered when there was nothing to catch, and that men and women did some of each, though both of them probably did a great deal more gathering than hunting. After all, the important thing was to get the day's food, not to define sex roles. . . . We presume too much when we try to read them in the scant record of our distant prehistoric past[24].

Hubbard closed her essay with a review of several feminist strategies that might clear androcentric science from evolutionary biology and the history of human evolution. The mythology of science suggested that these disciplines were objective; therefore, any bias would be filtered out over time. Hubbard argued:

> [This] might conceivably be so if scientists were women and men from all sorts of different cultural and social backgrounds who came to science with very different ideologies and interests. But since in fact they have been predominantly university-trained white males from privileged social backgrounds, the bias has been narrow and the product often revealed more about the investigator than about the subject being researched.[25]

She urged women who recognized androcentric myths to "do the necessary work in the field, in the laboratories, and in the libraries, and come up with ways of seeing the facts and of interpreting them."[26]

In their epilogue to *Women Look at Biology Looking at Women*, Hubbard, Henifin, and Fried wrote that one of the most liberating things that scientists can do for women is "to stop telling us what we are, while a male-dominated society sets the limits to what we can be."[27] Wilson might have been correct, they agreed, when he said in an interview that, "even with identical education and equal access to all professions, men are likely to play a disproportionate role in political life, business and science."[28] But, they continued, "This is not, as he would have us believe, because our evolutionary history has endowed women with domestic and nurturing genes and men with professional ones, but because conditions of work in the male-dominated professions do not suit the lives most women want or are able to live."[29]

In the acknowledgments, Hubbard thanked several of her colleagues— not only Rita Arditti and Stephen Jay Gould but also Garland Allen—for providing feedback on her manuscript of "Have Only Men Evolved?" before publication. Allen's mention is notable, as Hubbard had recently had a run-in with him in the pages of *Science for the People*. A social activist and a historian of the life sciences, Allen had written a letter to the editor complimenting the Sociobiology Study Group on its critiques,—particularly its coverage of "the overt sexism" in Wilson's book, in his talks and articles, and in the Canadian film. However, he wrote, the group had "persistently de-emphasized" another aspect of Wilson's writings, one that, in the long term, Allen believed, had the greatest potential for "political misuse. This is racism."[30]

In his view, it was important to acknowledge one important difference between racism and sexism: because the ruling class was composed of both sexes, sexism would probably never be employed "as a divisive tool" in the way that racism had been and was still being used. Allen argued that the profound divisions created by racism had always been greater than those caused by sexism. While both were components of colonialist, imperialist, and fascist nations, sexism had not formed the "central rationale" for any major historical developments. In contrast, innumerable "colonialist, imperialist, or fascist wars have been waged and justified on racist grounds."[31]

Given that Science for the People was just starting to build an institutional awareness of sexism, Hubbard, Arditti, and others were irritated by Allen's letter. Although he thought the study group was spending too much time on sexism in its critiques, most of the women in the group felt that members were not focusing on it enough. Arditti sent a reply to *Science for the People*: "Gar Allen's letter . . . throws me back eight years ago, when many people involved in social change movements failed to see the importance of the struggle against sexism." By creating a rank order—racism over sexism—he was suggesting, she said, that women's oppression was politically less important than racial oppression. In her view, this wasn't helpful. Allen didn't seem aware of the transferrable insights that had been developed by the feminist movement. Women's oppression was so old that it could offer a prototype for oppression. "Patriarchy existed well before capitalism," she pointed out.[32]

Hubbard also sent a letter to *Science for the People*. "I personally do not think that one can overplay the role of sexism in sociobiology," she wrote, noting that sexism "is at its very core." To her, Allen's letter showed a complete lack of understanding of the role of sexism. Regarding his comment about the ruling class, she responded that sexism offered women "satellite status" within the ruling class without granting them any power. She did not clarify, however, that even this marginal status within the ruling class was generally reserved for white women only. Certainly this was true in the United States.

In fact, though their arguments seemed oppositional, Hubbard, Arditti, Allen, and others were seeing sexism and racism as parallel, as operating separately. None of them were acknowledging that the two could work in tandem—for instance, in how they affected Black women. These activists hadn't fully integrated an analysis of how oppression was affecting Black women differently from white women. Their blindness on this issue reflected the blindness of the women's movement more broadly. Nor had *Science for the People* fully acknowledged that racism and sexism could reinforce and compound one another. There were no women of color taking part in the debate to reject the false binary.

Hubbard was angry with Allen's argument because it was coming from a male academic. Although he was an advocate for social justice, she did not see him as a feminist. Instead, she assumed that he would have more to lose if sexism were eliminated than if racism were erased because every facet of his family and professional life was structured by pervasive social sexism.[33] "Racism is basic to western capitalism and imperialism, but sexism has been with us since before the biblical patriarchs," she wrote. "As a feminist and a socialist I am ready to fight both, but I cannot fight alongside men who feel the need to prove that racism is the greater and more basic evil."[34]

Throughout 1977, letters about racism, sexism, and sociobiology raged back and forth in *Science for the People*. One focused on the priority of socialism and suggested that certain aspects of feminist reasoning could be "dangerous and divisive."[35] Another pointed to the history of abolitionists who had taken a stand against white women's suffrage. Yet another declared that, because racism was rooted in capitalism, defeating capitalism would end all racism—an argument that Black women would have contested, if they'd been given the chance.

Ruth Hubbard and Marian Lowe became so frustrated with the male members of the study group that they stopped attending meetings. Rita Arditti also stopped attending, largely because she'd had a falling out with Jon Beckwith. As Beckwith later recalled, "There was a debate. It was unpleasant. Most of the women left." They were exasperated, and they wanted to start something different, to create a place where their perspectives would be taken seriously.[36]

CHAPTER 8

MORE QUESTIONS FROM THE GENES AND GENDER COLLECTIVE

In January 1977, as Ruth Hubbard, Rita Arditti, and others were deciding they'd had enough of the Sociobiology Study Group, Hubbard traveled to New York City to attend a symposium titled "Genes and Gender," which had been organized by a group of women in response to E. O. Wilson's *Sociobiology*. Like Hubbard's colleagues in the Boston area, these women wanted to explode the myth that women's genetics determined their anatomy, physiology, behavior, and destiny. They, too, were concerned that the myth was dooming women to exploitation, limiting their choices, and preventing them from overcoming their oppression.

It was cold and windy in New York, and a major blizzard was making its way toward the city. Huddled in her warmest clothing, Hubbard walked up the broad stairs of the American Museum of Natural History and pushed through the big front doors. She was glad to step inside the grand rotunda, to feel warmth return to her face and hands, to step into the crowded meeting room. Organizers had planned for fifty participants, but more than 350 people had shown up, and the coffee ran out immediately. The ethnically diverse crowd included academics and students, parents with children. There was a $3 registration fee at the door; but those who couldn't afford it, didn't have to pay. The organizers wanted everyone to be able to attend.[1]

The symposium offered a number of different sessions—among them "Genetics for Understanding Gender," "Hormones and Gender," and "Society and Gender." After each presentation, small discussion groups grappled with the issues under discussion. One of the sessions that Hubbard attended was titled "Biology and Gender." The speaker was Dorothy Burnham, a molecular biologist, who had worked in New York City hospital labs for decades and also taught biology at Empire State College. In addition, she was a prominent civil rights activist and had been actively involved with the development of *Freedomways*, a 1960s-era journal of African American political and cultural life.

According to Burnham, it seemed irrational that some scientists were using genetics to support the "antediluvian ideologies of racism and sexism."[2] She explained how natural scientists had historically described the obvious physical differences between men and women and then moved on to concoct great theories about differences in intelligence, emotional makeup, and behavior. Scientists in the nineteenth century had emphasized the "genteel" nature of women, she said, claiming that their biology was "more suited to the parlor and the kitchen."[3]

Yet as Burnham pointed out, this argument neglected to mention poor and working-class women. Quoting the Black abolitionist Sojourner Truth, she looked out at her predominantly white audience: "Nobody helps me into carriages or over mud-puddles. Ain't I a woman? I have ploughed and planted, Ain't I a woman?"[4]

Hubbard listened carefully as Burnham spoke of the long days that working-class women and Black women spent laboring in factories or fields, before coming home to housework and childcare. Burnham pointed out the contradictions between eighteenth- and nineteenth-century beliefs about women's biology and destiny and the actual place that women held in society. This contradiction was nowhere more evident than in the treatment of Black women slaves. "The slave was treated like a nonhuman in every respect," she said. "In addition to physical cruelties inherent in the system, slave women were subjected to psychological and emotional traumas beyond belief." The qualities that supposedly differentiated women from men as a result of evolution and made them suited only to caring for homes and children "apparently did not apply to the Afro-American slave woman."[5]

Quoting *Sociobiology*, Burnham argued against Wilson's assertion that the division of labor between women and men was genetically based. As she pointed out, merely saying *that's the way it's always been, so it must be inherent* was not evidence for genetically based behavior. Yet, of course, the unspoken conclusion to such a statement was *that's the way it always will be*. Burnham also critiqued William Shockley, a Nobel Prize–winning physicist who had promoted the hereditarian idea of sterilizing women. As a result of such claims, women were too often the victims of involuntary sterilization, and a disproportionate number of them were Puerto Ricans, Native Americans, and other women of color.

Science and scientists had helped to justify the subordinate position of women, not only in society but within science itself. Burnham mentioned Rosalind Franklin's work, which had been essential to developing the model of DNA yet had gone unrecognized. Most women still never considered a career in science because they were discouraged from studying math and science. The hurdles for Black women were even higher and the opportunities "infinitesimally small."[6] "If I sound more like a Black feminist than a biologist," she said, "the reason is—that's what I was first. I truly believe racism and sexism interact and reinforce each other and the effort of both is not arithmetical but geometrical on the subject." Foreshadowing what Kimberlé Crenshaw would later call *intersectionality*, she declared, "I do not believe that anyone in America today would be given research money, facilities or publication privileges to prove a thesis relating to some supposed inferiority of white Anglo Saxon males."[7]

After the talk, Hubbard went up to speak to Burnham, and they discussed their common interests. Hubbard had recently responded to Garland Allen's letter in *Science for the People*, so the link between sexism and racism was on her mind. She asked Burnham if she'd be interested in contributing to a book she was putting together with Marian Lowe on the topic of women's nature. Burnham agreed to write a chapter about the vast difference between the realities of Black women's lives and the dominant myth of helplessness aimed at middle- and upper-class white women.

At the closing plenary of "Genes and Gender," organizers announced that they'd collected more money than they needed to cover expenses. Together, participants decided to use the extra funds to publish the proceedings. And thus, with the release of *Genes and Gender: On Hereditarianism and Women*, the Genes and Gender Collective was born.[8]

Hubbard wanted to stay involved, and she provided a link between the Genes and Gender Collective and Science for the People. Science for the People. In 1978, she took part in the AAAS conference in Washington, D.C.

Two scholars had organized a two-day symposium at the conference to bring together both advocates and critics of sociobiology.[9] The organizers let Science for the People and the study group use the same meeting room to hold their own meetings between the morning and afternoon sessions of the official symposium. For one panel, several women, many of whom had been part of the Sociobiology Study Group, presented papers. Hubbard and Lowe asked, "Can science prove the biological basis of sex differences in behavior?" Lila Leibowitz critiqued male dominance among primates. Freda Salzman considered the nature-nurture question as applied to humans. In addition, Ruth Bleier, a neurophysiologist from the University of Wisconsin, presented a paper on social and political bias in science, examining animal studies and noting how they were used to make generalizations about human behavior. Hubbard and Lowe eventually approached members of the Genes and Gender Collective, who decided to publish all the papers from the panel as *Genes and Gender II: Pitfalls in Research on Sex and Gender*. Edited by Hubbard and Lowe, it would be the second in what would become a series of publications.[10]

On the afternoon of the second day of the sociobiology symposium, both E. O. Wilson and Stephen Jay Gould were scheduled to speak on a panel, along with David Barash, a sociobiologist and psychologist at the University of Washington, and Eleanor Leacock, an anthropologist at the City College of New York, who was a critic of Wilson's and Barash's work. The hotel ballroom was packed with people who were keen to see the debate, especially the face-to-face exchange between Gould and Wilson, who would be the final speaker on the panel. But before Wilson could stand up to approach the podium, ten or so people rushed the stage. "Racist Wilson, you can't hide! We charge you with genocide!" someone shouted. Two people picked up a jug of water meant for the speakers, ran up behind Wilson, and poured it over his head. "Wilson, you are all wet!" they shouted before running off stage. The audience erupted in shouts, some against sociobiology but most of them aimed at the protesters.[11]

Finally, a session organizer moved to the microphone and apologized to Wilson, who stood up, shook off the water, and walked to the podium. The audience gave him a standing ovation. Before Wilson began to speak, Gould took the microphone and, to much applause, declared that such tactics were not the right way to debate the issues. Then Wilson, still dripping, delivered a talk in support of the genetic basis of human behavior. Afterward, Jon Beckwith, who was in the audience, stood up and verbally distanced the Sociobiology Study Group from the protestors. Clearly, attitudes toward protest action had shifted since the Science for the People demonstrations and the women's luncheon interruption at the 1969 AAAS convention in Boston.[12]

The Genes and Gender Collective wanted to keep alerting scholars and the public to the fallacies of human sociobiology. So, building on the publication of *Genes and Gender II*, it continued to hold regular conferences in New York City and publish their proceedings. Rather than depending on scientific journals, the collective released its work through Gordian Press, with the goal of reaching a broader audience and creating space for discussing the relationship between science and politics.[13] Hubbard, who was ready to leave the Sociobiology Study Group, wondered if she could start something similar in Boston, something that would focus on women. Dorothy Burnham's comments continued to resonate with her, and she wanted to do more to reach out to Black women in her own community.

So she began to organize a Genes and Gender Collective conference in Boston. Hubbard approached several women scientists, including Lowe, Leibowitz, and Anne Fausto-Sterling of Brown University, all of whom were concerned about sociobiology and the reappearance of theories of biological determinism. The intent was to proactively develop new areas of research that would focus on exploring and creating new forms of scientific knowledge. A working group met in February 1979 at Hubbard's office at Harvard.[14] Eventually, about thirty women, mostly local, agreed to participate in a seminar in early September. Titled "Beyond Determinist Thinking," the workshop would investigate the relationship among science and politics, human nature, and women's nature and would question science and objectivity.

Although Ruth Bleier lived in Wisconsin, she was invited to attend, and she was eager to take part. In her response she wrote, "I am looking forward to this conference both as an intellectual adventure and also as an exciting personal one. I think that we can not only provide new models for thinking about science, but we can also actually be a new way of intellectually interacting."[15] She pointed out, however, that "the cultural imperative of married heterosexism may be one of, if not the most powerful influences on a woman's development from childhood throughout life," affecting a woman's self-perception, aspirations, realization of potential, and control of destiny. She asked that someone present from a feminist-lesbian point of view, and she also asked, "Are any black women involved in the conference?"[16] This was an issue that the working group, made up entirely of white women, was already addressing. They invited several women scientists of color to participate, including Evelynn Hammonds, Dorothy Burnham, Beverly Smith of the Combahee River Collective, and Helen Rodriguez-Trias, whom Hubbard had met through her work at Boston City Hospital.[17]

By summer, a letter went out to all participants signed "the Boston Genes and Gender Group."[18] The draft program read:

> We need to make clear to each other our own assumptions about science and about social/political activity and change. In doing this, we must begin to explore the limitations in our own thinking. What are the results of our being members of a society whose values and science are those of a small dominant group made up of white men? We need to develop a sense of our own biases along racial, sexual, class and status lines if we are to develop our own significant new models.[19]

Hubbard asked participants to consider "the multiple overt and subtle connections that form the web of interweaving sexist, racist, classist and heterosexist oppressions that affect U.S. women" and the relationship of hereditarianism to these oppressions. She was particularly interested in the relationship between science and politics and asked the women to consider the following questions: "How do you think heredity and environment combine to affect human behavior?" and "do you think that the scientific method is applicable to the study of human behavior?" She also made the case that it is impossible to extricate nature from nurture in humans. Using a cooking metaphor, she asked, To what extent is custard the result of its ingredients? To what extent is it the result of the way in which those ingredients are mixed and cooked?[20] In an August letter to the working group, Bleier made a similar point, agreeing that there is no point at which something is genetically determined and immutable. Genetic expression, she argued, is influenced at conception and within the fetal environment and is in constant flux after birth. Nature and nurture are inextricably intertwined.[21]

In the late 1970s, the field of women's studies was growing, and gender and science had become part of that discussion. So in June 1979, as she was preparing for "Beyond Determinist Thinking," Hubbard took part in the first National Women's Studies Association meeting at the University of Kansas. Marian Lowe was chairing a panel on "Sex Differences and Sociobiology," and the panelists, who included Hubbard, and Bleier, discussed different aspects of biological determinism and the biological sciences as a political institution with bias. Hubbard had become an influential member of a growing community of feminists across the nation who were critiquing science.[22]

The gathering of the Boston Genes and Gender Group took place over three days at Currier House on the Radcliffe campus. While the initial idea had developed from concerns about sociobiology, the program ended up reflecting a much broader set of ideas. One session was titled "The Science and Politics of Determinist and Developmental Models." Others focused on reproductive technologies, the relationship between labor and biology, body image, intimacy, and sexuality.

Although organizers had tried to bring together a diverse group, the participants were mostly white and heterosexual.[23] Evelynn Hammonds, one of the few women of color present, listened quietly at the back of the conference room. She felt intimidated and wasn't comfortable about contributing to the discussion. Almost everyone else had academic or institutional positions, but Hammonds was still a graduate student. She wasn't a biologist, she was much younger than most of the other participants, and she was a Black woman. Still, she thought it was remarkable that Hubbard and others were willing to openly critique science in this way, especially in a region that was such a hub for science. She wanted to stay and listen.

Hammonds had recently had conversations with her young African American friends about the injustices that had been done to Black people in the South in the name of science. They wondered if, "in a Black Panther kind of mode," such a thing as Black science was possible. Could she become a scientist who could engage in work that would be helpful for Black people? As a student at MIT, Hammonds didn't know anyone else who had been asking those questions. So when she listened to these women scientists at Currier House, addressing the problems for women in science she was struck by their fierceness, their determination. They weren't addressing Black women specifically, but this was a start.[24]

Meanwhile, Hubbard was still wrestling with her own thoughts about diversity and inclusion. That year one of her students, Nancy Krieger, was preparing to serve as a teaching assistant in her "Biology and Women's Issues" course. In preparation, Krieger had convened a feminist study group that was planning to meet over the summer. In a letter, Krieger told Hubbard that she had reached out to "Third-world and Black women," hoping to diversify the white participants, but without much success. Two Asian American women had told her that "minority women" had too many other pressing issues to work on, especially regarding issues of racism. As time would show, they were right.[25]

CHAPTER 9

THE COMBAHEE RIVER COLLECTIVE AND THE MURDERS THAT SHOOK A CITY

In the 1970s, the Cambridge Women's Center was located on Pleasant Street in a residential neighborhood near Central Square. Founded in 1972, it was one of the oldest such centers in the country, and it provided space for activist organizing around issues such as birth control, abortion, rape crisis, and domestic violence. Evelynn Hammonds was one of the many women who spent time there. Usually, she attended a Wednesday evening discussion group, a welcome change after a long day in the lab. Those conversations centered around a range of topics, including Black women's history, working women, lesbian woman, women and violence, and women's health.[1]

The center was also the meeting place of the Boston chapter of the National Black Feminists Organization (NBFO). Its core group included Demita Frazier, a Chicago native who was doing community organizing in Boston; and Beverly Smith and Barbara Smith, twins from Cleveland. Barbara was teaching English at Emerson College in Boston, and Beverly was working on a public health master's degree at Yale but doing a field placement at Boston City Hospital. All three were civil right activists and would have a major influence on Hammonds.[2]

During NBFO meetings, Frazier and the Smith sisters discussed feminist issues facing Black women, how they had come to think of themselves as feminists, and what they were reading. The gatherings resembled the consciousness-raising meetings that Ruth Hubbard had once hosted in her living room. In fact, sometimes they met at Frazier's home in Dorchester, a Boston neighborhood with a large African American population, about six miles away from Cambridge. At the time both Dorchester and Cambridge had industrial and working-class areas. But while Cambridge was a center of elite, private institutions, Dorchester was a center of Black activism and the site of the public University of Massachusetts Boston. Martin Luther King Jr. had lived in Dorchester while he was completing his doctorate at Boston University, and Malcolm X had lived in nearby Roxbury for several years.

In the 1970s, Boston was a deeply segregated city that was in the midst of court-ordered school desegregation. Students from majority Black neighborhoods were being bused to majority white schools, and every day white students and parents were at schools and bus stops loudly protesting the court order. These scenes were frightening and often violent, and even people with no connection to the public schools were caught up in the atmosphere of fear and suspicion. Once, on her way to Frazier's house in Dorchester, Barbara Smith got lost and found herself in a hostile white neighborhood, and all she could think was "Uh oh, I really need to get out of here."[3]

In their meetings at the Cambridge Women's Center, the NBFO initially focused on abortion rights and issues surrounding sterilization. They started working with Dr Helen Rodriguez-Trias, a Puerto Rican–born pediatrician who ran the Committee to End Sterilization Abuse. Large numbers of Puerto Rican women, Native American women, and Black women, especially those receiving government support, were being sterilized without their consent. Rodriguez-Trias and her group staged demonstrations at Boston City Hospital and worked closely with lawmakers to push through legislation in Massachusetts to ban such abuse.

At the time, Beverly Smith was also working with the Women's Community Health Center in Cambridge; in fact, it became the subject of her master's thesis. Like the Boston Women's Health Book Collective (creator of *Our Bodies, Ourselves*), the center was woman-owned and -controlled, with a focus on educating women to know their own bodies. It went further, however, providing health care services, offering information on pelvic exams, dispensing contraception, and providing safe abortions.[4]

Beverly Smith hadn't intended to become a scientist, but her high school had required two years of science for college-bound students, so she'd taken biology and chemistry. Then, at the University of Chicago, she'd had to take physics, chemistry, biology, and mathematics as part of the general education requirement for all students, though her major was history.[5] Her interest in public health had grown out of her own life experiences. The twins' mother had died at age thirty-four, when the girls were just nine years old, so they were raised by their aunt, grandmother, and great-aunt. Then their great-aunt, who worked as a cook for a wealthy white family, died during the twins' first semester of college; and their aunt died only a few years later. Again and again, the Smith sisters were orphaned, and with each loss Beverly found herself drawn to issues of public health and Black women's health in particular..

At first NBFO felt like a good fit for Frazier and the Smith sisters. But over time they and their friends realized that their interests did not always overlap with the organization's. For instance, they wanted to talk about socialism and economic analysis, to open a vocal space for lesbians. They were also uncomfortable with NBFO's hierarchical structure; they preferred to work in a more collective style. So in mid-1976, their splinter group broke away from NBFO and became independent.[6]

The group needed a name. They considered Sojourner, after Sojourner Truth, a nineteenth-century abolitionist and women's rights activist. But that name was already the title of a local feminist newspaper. Then Barbara Smith had an idea. She had just read a biography of Harriet Tubman, which told the story of how Tubman had led a Union raid to liberate more than seven hundred slaves from plantations along the Combahee River in South Carolina. Why not name themselves after an action, not a person? The group agreed, and the Combahee River Collective was born.[7]

At the time, few Black women identified as feminists, often because they felt that doing so would divide the Black community. But the Combahee River Collective proved that there *were* Black women committed to feminism.[8] The terms *people of color* and *women of color* weren't in use during these years, so the Combahee River Collective gravitated to the notion of *third-world women* as a way of suggesting solidarity with the struggles of people around the developing world.[9] Combahee was wonderful, said Barbara Smith, because she could be all of herself in the same place:

I didn't have to leave my feminism outside the door to be accepted as I would in a conservative Black political context. I didn't have to leave my lesbianism outside. I didn't have to leave my race outside, as I might in an all-white-women's context.... That is what Combahee created, a place where we could be ourselves and where we were valued. A place without homophobia, a place without racism, a place without sexism.[10]

Shortly after the formation of Combahee, Zillah Eisenstein, a feminist academic at Ithaca College in New York, asked collective members to contribute to a book she was editing. Members decided to write a group statement that would clarify the source of their analysis. Frazier and the Smith sisters worked on multiple drafts amid long discussions. Frazier later recalled the moment when they discussed the notion that "we stand at the intersection where our identities are indivisible." The three talked about identity—that they weren't just female, just Black or lesbian, just working class, but embodied all of these identities at the same time. They were building their political views and actions based on that reality.[11]

In April 1977, the "Combahee River Collective Statement" was printed and distributed; and in 1978, it was included in Eisenstein's anthology *Capitalist Patriarchy and the Case for Socialist Feminism*. This was where Evelynn Hammonds, still a student at MIT, first read it. The statement made a deep impression on her.[12] She was struck by its point that separatism was not a viable strategy for Black women. Even if they were feminists and lesbians, they were in solidarity with progressive Black men so didn't advocate for the separatist approach that some feminists championed. Rather, "we struggle together with Black men against racism, while we also struggle with Black men about sexism."[13]

The statement also explicitly addressed racism in the women's movement:

As Black feminists we are made constantly and painfully aware of how little effort white women have made to understand and combat their racism, which requires among other things that they have a more than superficial comprehension of race, color and Black history and culture. Eliminating racism in the white women's movement is by definition work for white women to do but we will continue to speak to and demand accountability on this issue.[14]

While the statement made no reference to Wilson or sociobiology, it did engage with the topics of heredity, human nature, and biological determinism. The writers declared that biological determinism was a dangerous and reactionary basis upon which to build any political action. As one example, they noted that men are often socialized to act in oppressive ways, but the writers did not believe that biological maleness makes such actions inevitable.[15]

The statement quoted from a 1971 Black nationalist pamphlet that defined the male role as head of both household and nation because "his knowledge of the world is broader, his awareness is greater." According to the pamphlet, women were made "by nature" to function differently so they could not do the same things as men. The writers of the "Combahee River Collective Statement" disagreed, pointing out that feminism was calling these sex roles and power relationships into question.[16]

During these years, more and more feminists were discussing biological determinism. Given the centuries of racism and slavery in the United States, even people who had not followed the IQ debates linked to the work of Jensen and Herrnstein or the sociobiology debates linked to Wilson understood the concepts of inequality and inferiority. Members of the Black community were well aware that inferiority was often described as heritable. For hundreds of years, white people had told them that they were not fully human and that this was somehow rooted in their physical being and physical features. So the debates around determinism struck home.[17]

Even before drafting the "Combahee River Collective Statement," Barbara Smith had been arguing that feminism could activate a groundbreaking disruption in Black and third-world communities because it called into question the most basic assumptions about one's existence—in particular, the idea that biological and sexual identity determined everything. The irony, as she pointed out, was that among these communities "biological determinism is rejected and fought against when it is applied to race, but generally unquestioned when it applies to sex."[18]

So in July 1977, the Combahee River Collective started holding Black feminist retreats that gave participants the opportunity to discuss topics such as the theory and analysis of Black feminism, organizing skills and institution building, Black women's health, and the effects of isolation. The retreats attracted women beyond the regular members of the collective, including the inspirational Black lesbian poet, writer, and activist Audre Lorde. Women came from as far away as California, Chicago, and Washington, D.C. Some, such as Cheryl Clarke and Akasha Gloria Hull, went on to shape the growing area of Black women's studies. At the time there were few published books by Black feminists that included a gender analysis, but the collective members did their best to share information. At the events, Barbara Smith, who herself taught Black women's literature and Black women's studies, would set up a literature table where everyone could share mimeographed documents, and members began to discuss publishing as a tool for organizing. After a conversation with Audre Lorde, Barbara Smith started seriously thinking about establishing what would become the Kitchen Table Press.[19]

Although Evelynn Hammonds never attended a retreat, she was influenced by the dynamics of the Combahee River Collective, just as she'd been influenced by Shirley Malcom and "The Double Bind." It was exciting to talk about class and race and sexuality and power, to struggle with what it meant to be a radical, Black, lesbian feminist. For her, the collective was transformative.[20] And she was not alone: the Boston area changed because of the Combahee presence. Collective members were thoroughly woven into the politics, activism, and social relationships of the women's movement. Women across Boston and Cambridge now knew that there were Black feminists, and Combahee started to hold white women accountable for issues related to race and racism.[21]

Between January and May 1979, twelve Black women and one white woman were murdered in Boston, all within a two-mile radius in the predominantly Black neighborhoods of Roxbury, Dorchester, and the South End. Many were brutally strangled or stabbed, and some were raped before they were killed. Members of the Combahee River Collective were shocked by the murders, and their reactions had an impact on how the city responded.

At first, the *Boston Globe* did not give much space to the murders, reporting the discovery of the first two bodies on January 30 in a four-paragraph story buried on page 30. On January 31, an article about the third murder appeared on page 13. On February 6, the fourth murder in a week was covered at the bottom of page 1; and on February 22, the story of the fifth murder appeared in a small box at the bottom of the front page. On March 16, the sixth murder was covered on page 25. In contrast, the *Bay State Banner*, a Black community weekly, ran extensive coverage, starting on February 1 and continuing throughout the year.[22]

On April 1, several members of the Combahee River Collective attended a march to mourn the deaths of the six Black women who had been killed so far. They carried a banner that read, "Third World Women: We Cannot Live without Our Lives." The crowd of 1,500 people marched from Harriet Tubman's house in the South End, past the apartment of Daryl Ann Hargett, the fifth victim, on Wellington Street, and then to the Stride Rite shoe factory on

Lenox Street in Roxbury, where the bodies of the first two women, Christine
Ricketts and Andrea Foye, had been found.[23]

At the end of the march, a slate of speakers, most of them Black men,
stepped up to talk. They focused on the murders as racial crimes and highlighted the Black community's poor relationship with the overwhelmingly
white male police force and the rage of the white community around busing.
They declared, "We have to protect our women." Yet Barbara Smith, who was
listening, was appalled that none of them even mentioned violence against
women in their speeches. If this is all about race, she wondered, why are only
women being murdered? She and other members of Combahee understood
that some of the white feminists at the march did see the murders as a reflection of sexist violence against women. But the collective saw the murders as
both racist and sexist.[24]

Smith decided that Combahee needed to take action. She went home
to her apartment in Roxbury and started drafting a pamphlet, "Six Black
Women: Why Did They Die?" The next morning, over the phone, she read it
to her sister Beverly and other collective members. "Our sisters died because
they were women," she declared, "just as surely as they died because they
were Black." Refuting a phrase that had appeared in the *Boston Globe*, she
said, "It's no 'bizarre series of coincidences' that all six victims were female,"
and she pointed to statistics about the reality of violence against women in
the United States. The pamphlet included a sixteen-point action plan that
included safety strategies for women who were walking on the streets or traveling in a cab.[25]

The Combahee River Collective printed the pamphlet, but they kept having to reprint it and correct the subtitle because the number of murders kept
going up—from six to seven to eight to twelve. Eventually the Smith sisters
decided to just cross out the number by hand and insert the new one. Forty
thousand copies were distributed throughout the metropolitan area, and by
the end of the year seven Black men had been arrested for the murders of
eight of the Black women.

While none of the community activists were able to answer the title
question "Why Did They Die?" they were able to counter the mainstream
perception of Black women as "other" and bring greater attention to the
women whose lives had been destroyed. In this work, the Combahee River
Collective served as a bridge between feminist organizations and Black community organizations that did not consider themselves feminist.[26] During

one weekend in July 1979, the collective hosted two events to raise funds for the families of the women and build awareness around the murders. Both were poetry readings that featured the nationally renowned feminist poets Audre Lorde and Adrienne Rich as well as several local poets.[27]

The murders and the reactions to them influenced the feminist scientists across the river in Cambridge. Ruth Hubbard discussed them during her "Biology and Women's Issues" class, which by now had a hundred participants. There were lively conversations about the societal and institutional consequences of struggles around gender, race, and class.[28] Hubbard also reached out to Beverly Smith, inviting her to take part in the Boston Genes and Gender Group's "Beyond Determinist Thinking" seminar, but Smith was far too busy.

In early 1981, however, Hubbard reached out again to Smith to ask if she would contribute a chapter to a revised edition of *Women Look at Biology Looking at Women*. The first edition's authors had all been white middle-class women, and the co-editors wanted to include a broader representation. This time Smith agreed to take part, offering a chapter titled "Black Women's Health: Notes for a Course."[29] Hubbard also gave the new edition a new title: *Biological Woman: The Convenient Myth*. The preface acknowledged a need to honor diversity among women, and an expanded bibliography included resources from a wider range of sources.[30]

One of the new chapters, written by Helen Rodriguez-Trias, addressed sterilization abuse. She wrote, "It is shocking to many to learn that between 1907 and 1964 more than 63,000 people were sterilized under these eugenics laws in the United States and one of its colonies, Puerto Rico." She also pointed out a 1973 survey showing that 43 percent of the women who were sterilized in federally financed family planning programs were Black.[31]

Beverly Smith's chapter on Black women's health had been previously published that year in *All the Women Are White, All the Blacks Are Men, but Some of Us Are Brave*, an iconic anthology of Black women's writing co-edited by Barbara Smith. In it, Beverly Smith wrote, "By exploring this topic, we discover some of the ways in which living in this oppressive society breaks us physically and mentally and also how we have struggled and survived." The information was set up as a classroom aid, and the ideas and resources in the syllabus had been gleaned from Smith's own teaching curricula on women and health, her work on Black and third-world women's health, and her activism in the Combahee River Collective and the Committee to End Sterilization Abuse.[32]

The more diverse contributions to the second edition of Hubbard's co-edited book were proof that white women scientists had started to think about how Black women's lives, health, and biology might be affected by their circumstances. However, not all of the core feminist writings changed course so quickly. It took years, for instance, for the women involved with *Our Bodies, Ourselves* to fully embrace the experience of Black women. The book's first newsprint edition had been published in 1970, and by the third edition, in 1976, only one section of one page commented on "what it means to be a Black woman and a lesbian." In 1984, the preface acknowledged that, while the book "includes experiences and information gathered by women of color," it continued to be "a book written primarily by white women."[33]

In September 1979, Ruth Hubbard attended a large conference in New York City titled "The Second Sex—Thirty Years Later: A Commemorative Conference on Feminist Theory." The name nodded to the title of Simone de Beauvoir's *The Second Sex*, first published in France in 1949 and now a milestone in feminist philosophy. The conference offered more than thirty workshops on topics related to socialism, writing, violence against women, psychoanalysis, the sexual politics of pornography, women's nature, and biological determinism. Close to eight hundred women attended, including several scientists. Evelyn Fox Keller, the author of "The Anomaly of a Woman in Physics," was there, as was Sandra Harding, a feminist philosopher of science. Almost all of the attendees were white.

Robin Morgan, a founder of *Ms.* magazine, was scheduled to introduce the opening session, but first she gave the microphone to Susan McHenry, a Black feminist who worked with her at the magazine and was not listed on the program. In a brief speech, McHenry expressed outrage that so few Black women had been scheduled to speak. The program continued, but the issue that McHenry had raised hung heavy in the air.[34]

Audre Lorde was one of a handful of Black women whose name appeared on the program. Like De Beauvoir, Lorde rejected the notion that biology was the basis for women's social role in the world, believing instead that their material conditions were more fundamental. Hubbard certainly agreed and was eager to hear her speak.

At the last formal session of the conference, Lorde stepped to the podium and condemned the organizers for the limited range of speakers. Why hadn't they attracted a larger number of women of color? Why was she seen as the only possible source for names of Black feminists? Lorde had agreed to take part in the conference because she thought she would be commenting on papers about the lives of American women who were dealing with race, sexuality, class, and age. It was "academic arrogance," she said, to think that any discussion of feminist theory could happen without input from poor women, Black and third-world women, and lesbians.[35] Lorde quoted Adrienne Rich to remind the audience that white feminists had educated themselves about an enormous number of issues during the past decade. Why hadn't they educated themselves about Black women and the differences in their life experiences, especially since this knowledge was so important to the survival of the women's movement?[36]

Lorde challenged the participants to consider that the conference was employing the same tools of oppression that the participants abhorred in patriarchy: "Those of us who stand outside the circle of this society's definition of acceptable women, those of us who have been forged in the crucibles of difference—those of us who are poor, who are lesbians, who are Black, who are older—know that survival is not an academic skill. . . . For the master's tools will never dismantle the master's house."[37]

For some of the women in attendance, Lorde's message rang true. But the hard reality was that white feminist academics depended on white male approval for success, promotion, and tenure. Moreover, even as they tried, as feminists, to distance themselves from the patriarchy, they increasingly held power themselves, especially in women's studies programs. Lorde's reference to "the master's tools" would become a ringing metaphor as feminists sought new ways to interact with organizations, institutions, and society.

Yet some women scientists were skeptical, and Evelyn Fox Keller was among them.

CHAPTER 10

EXPLORING THE GENDER OF SCIENCE

Ironically, the most significant obstacle that women have long faced in science is not their own nature but the widespread social myth that science is somehow inherently masculine. This myth presents objectivity, reason, and mind as male; subjectivity, feeling, and nature as female. Where did this belief come from? And what is it doing in science?[1]

These two linked questions shaped Evelyn Fox Keller's research and writing for more than fifteen years. She was not deeply involved in the sociobiology debate that erupted in the 1970s after the publication of Wilson's book. In part, that's because she had been teaching at the State University of New York at Purchase so was not engaged with Boston-area organizations such as the Sociobiology Study Group and Science for the People. Nor was she involved with the Genes and Gender Collective in New York City, though she did participate in feminist consciousness-raising groups in Purchase, about thirty miles north of the city. In addition, the growing field of Black feminist scholarship was not having a significant impact on her thinking as a feminist.

Yet Keller did see that questions about sex differences were becoming fraught in the late 1970s and early 1980s. She recognized, as most scientists did, that both physiology and environmental influences play a role in human development and behavior. To what extent does one of these influences play a larger role than the other? And how can that question be answered scientifically?

Then, after twenty years away from Boston, Keller moved back to the area in the fall of 1979 to take up a year-long fellowship in MIT's new science, technology, and society program. That November, she was one of the speakers on a panel at Harvard that had been organized by the Sociobiology Study Group. Freda Salzman, also a panelist, spoke about sociobiology and sex roles, while Keller criticized a new study claiming to illustrate sex differences in math ability.[2]

Unlike Ruth Hubbard, the Genes and Gender Collective, and the Sociobiology Study Group, Keller was not interested in how scientists portray women's biology. She was much more interested in how the social concept of gender—that is, what is regarded as masculine or feminine—shapes the way in which we think about science. But she struggled to make her focus clear to her friends and colleagues. A person might ask, "What are you working on right now?"

"Gender and science," she would answer.

"How interesting!" the person would say. "And what are your findings about women and science?"

Keller was frustrated by such exchanges. Again and again, she would explain that gender related to both women and men. She wasn't thinking about women per se but about culturally entrenched beliefs around masculinity. Yet biologists were often confused by the term *gender*. They tended to see it as the biologically determined state of being male or female, whereas Keller was talking about cultural influences. The term *beliefs* was similarly misinterpreted. When Keller spoke of popular beliefs about science (many of which were inaccurate), people often thought she was talking about her own perceptions.[3]

As Keller began to explore the association between the concept of being masculine and the concept of being scientific and objective, she noticed that many academics were hesitant to discuss the topic. She wondered how scholars who focused on the philosophy and sociology of science had failed to see that it required analysis. She acknowledged that feminists such as Arditti, Hubbard, Salzman, and others were addressing the issue, but the virtual silence of the rest of the academic community was telling. In Keller's view, the association between the masculine and the scientific was an unexamined social myth—one that was seen as both self-evident and nonsensical. This association was considered to be common knowledge, yet few people wanted to point out its odd conflict with the idea of science as neutral and objective.[4]

Keller's 1978 article "Gender and Science" made exactly this argument: that scientists were hesitant to talk about the historical association between being scientific and being masculine because many people assumed that science was objective. The association of masculinity with scientific thought had taken on the status of myth, one that was widely accepted within science and among the broader public. Yet this association clashed with the accepted idea that science was also objective. In other words, how can science be objective and be masculine at the same time? Keller pointed out the confusing historic association between being masculine and being objective. These were cultural stereotypes, and she believed that one consequence of them was the fact that an overwhelming number of scientists were male.[5]

Once it had been common for scientists, parents, and teachers to declare that women didn't have the skills or the brains to undertake a scientific career. Keller felt that second-wave feminism had made it less fashionable to state this claim bluntly, but the underlying myth still existed. She pointed out the kinds of language used to describe science:

> When we dub the objective sciences "hard" as opposed to the softer (that is more subjective) branches of knowledge, we implicitly invoke a sexual metaphor in which "hard" is of course masculine and "soft" feminine. Quite generally, facts are "hard," and feelings are "soft." A woman thinking scientifically or objectively is thinking "like a man"; conversely, a man pursuing a non-rational, non-scientific argument is arguing "like a woman."[6]

Keller didn't explore the relationship between science and white supremacy. She was examining gender only, without considering the impact of other issues such as racism, militarism, or capitalism. What she wanted to convey was that society held beliefs about differences between the sexes that far exceeded anything that could be attributed to hormones or biology and that these ideas were driven by powerful cultural and psychological forces. In her article, Keller tried to make it clear that what she was discussing was a system of beliefs about *masculine* and *feminine* rather than any real, intrinsic differences between male and female. Nonetheless, in the coming years, her discussion of this topic would consistently be misinterpreted.

Both Keller and Hubbard attended "The Second Sex" conference in New York City, and Keller had prepared a presentation titled "Nature as 'Her.'" In it, she put forward a question about the relationship between nature and gender. As Arditti had done in her own writing, Keller examined the work of Francis Bacon, sometimes credited as the architect of modern science. In the late 1500s and early 1600s, he had described the purpose of science as "the control and domination of nature" and said that he viewed the relationship between man and nature as "a chaste and lawful marriage." Keller argued that Bacon's metaphor had come to deeply influence human conceptions of nature. Thus, as modern science developed, nature was seen not only as female but specifically as a wife over whom science has control and power. Profoundly, Keller asked why, "for science, the association of women and nature invites domination, while for the poets, it invites release, escape, transcendence?"[7]

At the time Hubbard and Keller were taking very different approaches to the study of science and objectivity. For the first time in her career, Hubbard had stepped back from her work in the lab to learn from non-scientists such as Audre Lorde, Beverly Smith, and other Black feminists and friends. She was embracing new knowledge. Keller, however, was unsympathetic. She did not see that someone's perspective or standpoint was useful in discussions of science. She didn't question the role of scientific enquiry and observation.[8]

This disparity in views erupted during "The Second Sex" conference, when Keller and Hubbard had a heated argument about the issue of who decides what counts as science. Though Keller was twelve years younger than Hubbard, the two had known each other since the 1950s, when Keller had been a graduate student in physics at Harvard and Hubbard was a young biologist in the lab. Now Hubbard wanted to know if Keller thought that women and feminists should accept the male establishment's definition of science. Hubbard thought this was an important question to explore while Keller thought it was ludicrous. This was the first big blow-up between the women, and both stood their ground. In essence, they were arguing about the definition of knowledge and the ways in which one could access knowledge. By now, Hubbard had started to question the reductionist approach to science, and Lorde's metaphor of "the master's tools" had resonated with her. Yet even though Keller, too, had begun to question the objectivity of science, especially in relation to gender, she was not questioning other fundamental tools of science as a discipline. She did not want to create new tools. She was a scientist and proud of it.[9]

In truth, neither Keller nor Hubbard had fully embraced Lorde's points about race, class, and difference. The impasse between them was influenced by their focus on two very different topics. Hubbard was interested predominantly in the science of gender: that is, in how biologists, medical professionals, and other scientists presented gender. Keller, on the other hand, wasn't interested in the science of gender. She was interested in the gender of science.

In 1977, Keller watched a film about Antonia Brico, who, in 1930, had been the first woman to conduct the Berlin Philharmonic Orchestra. She found the film powerful, moving, and inspiring. But as she walked home afterward, she realized that she had never seen a film about the struggles and aspirations of a woman scientist, with the one exception of Marie Curie. She couldn't think of a single novel with a woman scientist as a main character. Why? Certainly, the lives of women scientists could provide dramatic material for literature and film, but apparently no one had chosen them as a focus for creative work.[10]

As Keller walked into her apartment, a ringing telephone interrupted her thoughts. When she answered, an unfamiliar male voice asked, "Hello, is this Evelyn Fox Keller?"

"Yes," replied Evelyn.

"I've just read your article 'The Anomaly of Women in Physics,'" the man said, "and I think you should write something about Barbara McClintock."[11]

Keller suddenly remembered her glimpses of McClintock at Cold Spring Harbor in 1960, during her summer there as graduate student. She recalled being terrified of McClintock because she seemed to be so alone, because she confirmed Keller's worst fears about becoming a woman physicist. Back then Keller hadn't wanted to get to know her.

Now, twenty years later, she felt differently. She had just been reflecting on how difficult it was for women to forge unconventional paths, how few of them had received public attention. The more she thought about it, the more she believed that McClintock's life experiences must be extremely interesting. Someone should write about her, should make a film about her. Perhaps, Keller thought, she herself could write a short article about McClintock, publish it, and then convince someone else to make the film.

Keller called a mutual friend, who told McClintock that Keller was interested in interviewing her and sent McClintock a copy of "The Anomaly of a Woman in Physics." McClintock read the essay and agreed to meet.[12] So in the fall of 1978, Keller drove to Cold Spring Harbor with her tape recorder. Along the way, she mulled over the different ways in which McClintock had been described to her: "intimidating," "difficult to approach," "a very private person."[13]

But when they actually met, McClintock greeted Keller warmly and made her comfortable in a large armchair. The two immediately began talking about science. Keller later recalled that McClintock stayed in control of the meeting, asking Keller about her background, her interests, and her motivations for wanting to write this article. Only then did she agree to let Keller ask questions, and they spent the next five hours in deep conversation. On her way home, Keller thought, "My God, I have just encountered—for the first time in my life—a 'great mind' in a woman's body."[14] Her training in theoretical physics had taught her to appreciate great minds, yet all the ones she had read about or been told about were men. Until McClintock, she had not come across one single great mind that belonged to a woman. This was reason enough to write about her, Keller thought. Interestingly, however, at this juncture she didn't critique her own association of great minds with men.[15]

Nonetheless, Keller was conflicted about the project. She had already started to write about issues of gender and science and the impact that ideologies about masculinity and femininity were having on the production of science, and she worried that writing about McClintock would distract her. But at some point she decided that she needed to establish the right of a woman to be a great scientist. She set aside her gender and science research to work on a full-length biography of McClintock.

Not all of Keller's friends were sympathetic. "She's not a feminist," said some of them. "Why are you writing about her?" Others saw her as an eccentric who "has devoted her entire life to her work."[16] As Keller took into consideration these different views, she realized that she wanted to establish McClintock's right to be different from any female stereotype, whether feminine or feminist. "I was troubled by the rejection of this as a feminist project by some of my most committed feminist friends," she later said.[17]

After three years of research and writing, Keller sent the manuscript to the publisher W. H. Freeman. For six months press staff sat on the project, asking for changes. They didn't like the title, *A Feeling for the Organism* (a quote from McClintock), and they wanted to edit the book extensively. Keller refused to agree to these changes, and the publisher continued to stall. Eventually, Freeman did agree to release the biography, and it appeared in 1983 in a much smaller print run than had been originally promised. Yet despite the publisher's lack of enthusiasm, the book sold out within a few weeks.[18]

Keller felt that she'd achieved her goal: no one, neither feminists nor scientists, saw the book as a feminist manifesto or an account of feminine science.

Rather, it was a book about a scientist set within the well-established tradition of great minds. Keller insisted that the book's title did not imply that *feeling* was an especially female trait. It was a human trait, accessible to all. The book, she explained, was the biography of both an individual and a science—genetics. It was about the nature of scientific knowledge and the interaction between individual scientists and the broader scientific community.[19]

Barbara McClintock was born in 1902, two years after Gregor Mendel's early work on the biological inheritance had been rediscovered. At the time, no one was sure how inherited traits were transmitted from one generation to another. The term *genetics* did not exist until 1905, and the word *gene* wasn't recognized until 1909. Thus, McClintock's learning and career was entwined with the development of the field.[20]

In 1920, the number of women trained in science was higher than it had ever been in the United States. At Cornell University, 203 graduates earned a bachelor of science degree in 1923, and seventy-four of them—including McClintock—were women. Fifty years later, in 1970, women graduates in science disciplines had dropped to roughly half that number.[21] For Keller's contemporaries, the disparity was startling.

McClintock studied botany at Cornell and got her PhD there in 1927. She explored cytogenetics—the study of the structure and function of chromosomes—and established herself as a leading maize, or corn, geneticist. In 1939, she was elected vice president of the Genetics Society of America, and she became a member of the National Academy of Sciences in 1944 and the president of the Genetics Society in 1945.[22]

In the early 1950s, McClintock proposed the concept of *transposition*—that genetic elements move from one chromosomal site to another carrying instructions to the cell. At the time, genes were thought of as simple units in a fixed, linear sequence. They were compared to beads on the string that was the chromosome. It seemed unlikely that genes could move from one site to another or from one chromosome to another. But McClintock was proposing a more fluid, dynamic process. She was suggesting that the instructions in a cell are interactive rather than linear. Most other scientists in the field found McClintock's ideas improbable, and she spent twenty-five years unsuccessfully trying to change their minds.[23]

In the mid-1950s, molecular biology took the world by storm. In this new environment, McClintock's manner of working seemed old-fashioned, and her ideas didn't quite fit into the new direction of genetics, with its focus on a reductionist approach. Researchers such as Watson and Crick were looking at the smallest building blocks that they believed could explain the complex forms of completed organisms. Soon after announcing their discovery of the double helix structure, Crick declared that the "central dogma" of the new genetics was that DNA makes RNA makes protein in a one-directional flow of information.[24]

McClintock's concept of transposition was not welcomed by her peers. Writing about the nature of scientific knowledge, Keller explained that it grows from the interaction between individual ideas and the group dynamics of the scientific community. Sometimes, however, "the interaction miscarries and an estrangement occurs between individual and community." She asked, "What role do interests, individual and collective, play in the evolution of scientific knowledge?" In other words, when people ask different questions and use different methodologies, how does this affect communications among scientists?[25]

Only in the acknowledgments did Keller refer to her own work on gender and science. Writing the book "sharpened my thinking" on this topic, she said. "In her adamant rejection of female stereotypes, McClintock poses a challenge to any simple notions of a 'feminine' science. Her pursuit of a life in which 'the matter of gender drops away' provides us instead with a glimpse of what a 'gender-free' science might look like." Notably, Keller also thanked Ruth Hubbard for reading the manuscript before publication.[26]

When Keller first spoke with McClintock in 1978, the seventy-six-year-old's career was on the cusp of change. Something had shifted, and her work was finally being recognized. In 1979, she received two honorary degrees, one from Rockefeller University, the other from Harvard. In 1981, she was named a MacArthur fellow in the first year of the program's existence, winning $500,000. That same year, she received the Lasker Award for Basic Medical Research, and numerous other awards followed.[27] Then, in December 1983, five months after *A Feeling for the Organism* was published, McClintock won the Nobel Prize.

After McClintock won the Nobel, Keller's biography became a hot commodity. The Nobel committee couldn't find any copies of the book because it had sold out. Eventually it was translated into ten languages, and it made Keller famous among a broad audience of readers, though that didn't necessarily translate to prestige within the academy. Moreover, Keller still had a question: why had it taken so long for other scientists to recognize McClintock's accomplishments?

She later mused, "Some feminists who had earlier dismissed McClintock for doing science 'just like a man,' and had rejected the writing of her story as 'not a feminist project,' now sought to embrace her as a feminist heroine."[28] These readers were misreading Keller's intentions. Just as people had misunderstood the terms *gender* and *belief* in her research, they were now misunderstanding the purpose of her biography. Many were focused on the notion of "feminist science," assuming that it was a type of work that women scientists would naturally do. This was not at all what Keller was trying to say. For her, and for other scientists such as Hubbard and Arditti, the idea that there was *anything* that women would naturally do was very problematic. Keller knew that it was terribly damaging to suggest that women's science was different from men's. Women scientists already had plenty of barriers. No one needed to add another one to the list.[29]

Keller had another concern. Even as feminists were seeing McClintock in a new light, the research scientists who had previously dismissed her now wanted to claim her as one of their own. This is a common dynamic. Women's ideas, especially those of women of color, are often initially rejected. But when these ideas gain traction in the field, they are co-opted by the people who originally rejected them.[30]

While working on *A Feeling for the Organism*, Keller had extended her fellowship time at MIT, and in 1981 Northeastern University hired her as a professor of mathematics and humanities. There she returned to her earlier project and in 1985 published a series of essays under the title *Reflections on Gender and Science*. In one of those essays she clarified her views on the various reactions to her biography of McClintock: "In a science constructed around the naming of object (nature) as female and the parallel naming of subject (mind) as male, any scientist who happens to be a woman is confronted with a contradiction in terms." Keller suggested that both nature and mind needed to be reconceived as gender neutral and that science should be seen as a human endeavor rather than a male one.[31] Again, however, a number of readers misinterpreted her intent. Many feminists and scientists were frustrated by her, others were inspired, and still others avoided her.

This pattern continued for much of the rest of Keller's career. On one occasion, during a daylong workshop at MIT that featured a group of women scientists talking about gender and science, she tried to make the case that men as well as women are shaped by gender ideology. In her talk, she argued that the term *feminine* could be reconceived to describe the human traits of both men and women. Afterward, one of the scientists wrote to her, saying, "That was really interesting, but I still don't believe that women's intuition is inscribed in their chromosomes." Keller was still failing to get her point across.[32]

In 1982, before publishing either of her books, Keller wrote an article, "Feminism and Science," that posited a way of thinking about three different aspects of the growing feminist critique of science. First was "the liberal critique"— the idea that women were facing unfair discrimination in the field of science, which was linked to a call to promote equal opportunity for women in these disciplines. Margaret Rossiter's work fell into this category, as did some of the work coming out of the National Science Foundation to encourage women to enter the sciences. Such approaches were having an effect as, starting in the mid-1980s, women in science became an enormous field, with committees, institutional infrastructure, and large development budgets.[33]

The second, somewhat more radical critique recognized that most of the people who had been doing science during the past several centuries were white men who had brought a particular bias to their work. Keller pointed out that this problem was particularly important in medicine, the health sciences, and biology. Hubbard's and Arditti's work fell in this category, as did Anne Fausto-Sterling's.

The third, most radical critique argued that the design and interpretation of scientific experiments were biased and that this required people to rethink the methodology of science and the assumption that science was objective. In the 1970s, the debates within Science for the People, the Sociobiology Study Group, and the Genes and Gender Collective fell into this category, as did the work of feminist scientists and philosophers of science such as Sandra Harding.

These three feminist challenges to science would play out very differently from one another in the 1980s.[34]

CHAPTER II

THE RELUCTANT FEMINIST

In June 1976, the *Radcliffe Quarterly* published "The High Price of Success in Science" in which Nancy Hopkins, a biologist, disputed the idea that a woman could be a successful wife and mother as well as a successful scientist.[1] A native of New York City, Hopkins had completed her PhD at Harvard and was now an associate professor at the Center for Cancer Research in the biology department at MIT. She was much younger than Ruth Hubbard, Rita Arditti, and Evelyn Fox Keller; and, in her view, all of them were experiencing and critiquing the science world in ways that were less than useful for young women early in their careers. Young women couldn't change the system, she argued, so they must focus on the choices they had to make to become part of it. Hopkins agreed with the older cohort that the intellectual work done in the laboratory came as naturally to women as it did to men. The problem was that the professional work of science—teaching, supervising, writing grant proposals, producing journal articles to get a job—had been constructed by men, and women did not fit well within this framework. As she was working on the article, Hopkins asked many of her male colleagues if they could have become successful scientists if they were women. Each one responded no.[2]

Hopkins had graduated from a private New York City high school in the 1950s and had assumed that she was destined to marry and have children before she turned thirty. Nonetheless, by the time she was an undergraduate, Harvard and Radcliffe students were taking classes together, and received Harvard diplomas. Her mother hoped she would marry a Harvard man and thought a college degree would help her to find a job if he were to die. But Hopkins took her educational chance seriously.

She began to consider going into medicine and knew she needed a biology course. So she signed up for an introductory class with James Watson, the researcher who, with Francis Crick, had won a Nobel for identifying the helical structure of DNA. In the class, Watson spoke about DNA and the genetic code that was still being unraveled, suggesting that everything biological was in some way the product of this one molecule and its nucleotide sequence.[3] Forget medicine, thought Nancy. I want to be a molecular biologist. She floated out of the classroom, convinced she'd found the meaning of life.[4]

Toward the end of the semester, Hopkins asked Watson if she could work in his lab, and he said yes. Hanging around the lab, soaking in the atmosphere of excitement, made her feel like a "science groupie." During lunches with Watson, she and other students were privy to the latest science news, hot off the press. In this heady atmosphere, Hopkins wasn't thinking about getting a PhD or becoming a professor but about making a discovery and winning a Nobel.[5]

Yet she also had a serious boyfriend, and soon Hopkins started to wonder if science and motherhood were compatible. As her male colleagues spent up to seventy hours a week in the lab, their wives stayed home and took care of the children or worked in less demanding jobs. Already, Hopkins was feeling the pressure of time. She decided that she had to do as many experiments as possible throughout her twenties. Then, just before she turned thirty, she would have children. In an era without amniocentesis or legal abortion, she believed she had to make a plan.[6]

One day in 1964, as Hopkins was leaning over a bench in Watson's lab, she heard the door open. A large, older man burst through the door, brimming with cheer. She thought she recognized him; but before she could think, he bounded across the room, settled his hands on her breasts, and looked over her shoulder as if he were reading her lab notebook. Shocked, she quickly moved away and turned around. The man was Francis Crick. Her first thought was, How do I get out of this situation without embarrassing him? He was such a famous person. In 1964, the term *sexual harassment* did not yet exist, and Hopkins didn't know how to handle the situation.[7] Nor could she avoid Crick. That night Watson would be giving a party for his research partner, and Hopkins knew she was expected to attend.

The party was crowded. Senior Harvard faculty were there, along with

Watson's students, and everyone drank the strong punch. After the party, Watson and Crick escorted Hopkins back to her dorm on the Radcliffe Quad. The men talked and laughed as if everything were normal. So Hopkins said nothing.[8]

Still, she kept her eyes open, and she noticed that very few women were becoming independent scientists. Even those with a PhD often dropped out of their chosen fields, "unless they married a powerful scientist or were otherwise able to get a long-term position in a professor's lab." Hopkins herself didn't marry a scientist (her eventual husband, Brooke Hopkins, had been an English major), but she had a powerful mentor: Watson. Even he, however, spoke of her condescendingly, describing her in a recommendation letter as "a bright and extremely pleasant Radcliffe girl."[9]

Despite her observations and experiences, Hopkins didn't think that sex discrimination was causing her any problems, and she wondered what the feminists were complaining about. She wanted nothing to do with them; she only wanted to focus on her science.[10] Yet a part of her realized she was unusual. When she asked another young scientist why Watson had chosen to mentor her, he explained that everyone was curious to see if a woman could make it in science, and Watson thought that maybe it would be her.[11]

In "The High Price of Success in Science," Hopkins described the usual route of a young scientist. A graduate student earns a PhD in their mid-twenties, and their scientific reputation is based on productivity and their publications so they can get a good job. At this point, however, young women scientists often choose marriage and family, stepping out of their discipline just as men are intensifying their efforts. In the lab, there is enormous pressure to succeed and much competition. Research is a full-time job, as is university teaching, as is raising children. It was hardly surprising, then, that few scientists were women. Hopkins wrote of encountering the "occasional survivor," but such women had hardly ever followed the traditional route. Rather, they'd married a scientist and worked in his lab (like Ruth Hubbard), or they'd worked in a research institute rather than a university (like Barbara McClintock).[12]

Hopkins had met McClintock at Cold Spring Harbor in 1972 and had noticed her isolation from the new wave of molecular biologists. During their conversation, McClintock had urged Hopkins not to take a job at a university because she wouldn't be able to stand the discrimination. At the time Hopkins had wondered what McClintock was talking about. She was convinced that her generation was different, that she wouldn't have a problem.[13] But her 1976 article revealed that discrimination against women was still rampant, and Hopkins found "the idea of trying to cope with one's personal life while being a woman scientist" even more horrifying. She asked a small sample of women to tell her how they had managed. Their answer? "[By] not having children and usually by not even being married."[14]

A few months after Hopkins's article appeared, Ruth Hubbard published a review of a new biography of Rosalind Franklin, written by Anne Sayre. As Hubbard noted, many women scientists have felt forced to choose between professional life and marriage, and Franklin was one of them. But the divide often doesn't end there. Most scientific laboratories are structured like a "patriarchal household," and by means of this structure Watson "obliterated Franklin, the scientist." In *The Double Helix* (1962), he described "Rosy" as the "stereotype of the unattractive, dowdy, rigid, aggressive, overbearing" woman—a portrait that was not recognizable as Rosalind Franklin. Throughout his book, he made numerous sexist references to the role of women both in and outside the laboratory and his own need to retain power. One remark that Hubbard highlighted was "for the first time, I had a real incentive to learn some crystallography: I did not want Rosy to speak over my head."[15]

Hopkins had read *The Double Helix* in 1968, but she did not read Sayre's biography or Hubbard's review until the mid-1980s. Even at that late date, the book didn't impel her to join the feminist critiques of science.[16] She was still focusing on women's struggle to fit into the existing system. In 1976, Margaret Horton Weiler, an assistant professor at MIT, said that Hopkins' *Radcliffe Quarterly* article had "done a disservice to women and to science." "I find it especially difficult to understand her attitude because I too am a scientist at MIT (in physics). . . . Women have a long way to go in science as in other walks of life. . . . There aren't enough women at the top—too many of them listen to arguments like Nancy Hopkins."[17]

Hopkins had thought she was stating the obvious. It had never occurred to her that anyone would disagree with what she saw as a biological problem: childbearing years conflicting with career growth. Only much later did she begin to think about the situation as institutionalized discrimination. Who would design a profession in this way if you were designing it for people who had children?

In 1973, Hopkins's husband, by then an assistant professor of English at Harvard, decided to end the marriage. This took her by surprise. She'd loved that his work was so different from her own, but she hadn't believed he felt the same about her work, which was a problem. They'd never talked about science.[18]

Without her husband, Hopkins felt very alone. Her father had died while she was in college, her mother died of cancer soon after her divorce, and her sister moved to Singapore. Her plan of doing science until she had children no longer applied. After much thought, she decided she wouldn't remarry. She wouldn't have children. She would be a scientist for the rest of her life.[19]

Hopkins had been offered two faculty positions, one at MIT, the other at Harvard Medical School. Watson advised her to take the MIT job, where she could work with Nobelists David Baltimore and Salvador Luria at the newly constructed Center for Cancer Research. The timing was perfect, she recalled, because cancer was her chosen field of research.[20] As a child, she'd been terrified when her mother was diagnosed with a mild form of skin cancer. In those days, no one had dared to mention the word, but Hopkins always remembered her own fear. So she'd decided, after earning her PhD in 1971, to study the genetics of tumor viruses. Gradually, she'd shifted from research on DNA viruses to RNA viruses, which were then seen as a likely cause of many human cancers. The new center had an entire floor devoted to her field of study.[21]

With this new opportunity before her, Hopkins assumed that gender discrimination would become a thing of the past. The 1964 Civil Rights Act had made discrimination illegal, and the affirmative action laws passed in the early 1970s meant that universities were required to hire women or lose their federal funding. Hopkins was excited about becoming a science professor at MIT and devoting her life to research. But reality was different from her expectations. The environment at the center was difficult, much more difficult than it had been in Watson's lab. At MIT she was surrounded by wonderful scientists, but she also was on her own. Like many other women scientists, she knew that something was wrong but couldn't identify what it was.[22]

CHAPTER 12

THE WELLESLEY SEMINAR

In 1980, after Evelynn Hammonds completed her master's degree in physics at MIT, she took a job as a computer programmer at the Polaroid Corporation in Cambridge. Polaroid was a high-tech company (the *Boston Globe* later described it as the Apple of its time) headquartered in an art deco building along the Charles River. But each morning, as Hammonds drove past the MIT campus on her way to work, her heart sank. She felt she had failed at physics.[1]

Hammonds's thesis had focused on solid-state experimental physics: specifically, an X-ray scattering study of two-dimensional systems. Yet despite receiving recognition for her work, she had decided not to pursue a PhD. Her experience at MIT had left her feeling completely exhausted and defeated. She wanted to do physics, but she didn't know how she could survive another few years in that environment. Years later, when someone asked why she hadn't transferred to another institution to study physics, her answer was simple: it had never crossed her mind. She had convinced herself that MIT was the best place in the world to do physics. If she couldn't succeed there, then she had failed.[2]

While she was working as a computer engineer at Polaroid, Hammonds got to know several members of the Combahee River Collective, including Demita Frazier and Barbara and Beverly Smith. She'd read Beverly Smith's chapter on Black women's health, and now she was spending many evenings with her. Smith was serving as a mentor to young Black women in Cambridge, and Hammonds and the others soaked up her commentary about the reality and politics of being a Black woman in the United States.[3]

Soon Hammonds began writing articles and essays for *Sojourner*. Then, after the publication of *But Some of Us Are Brave* in 1982, Cheryl Clarke, who had also been a member of Combahee, invited her to join a public conversation about Black women writers. The group discussed how difficult it was to write and be heard as a Black woman, and in 1983 a transcript of the talk was published in *Conditions*, which Clarke described as an antiracist lesbian journal.

Hammonds enjoyed this sense of community with other Black women and continued to go to events at the Cambridge Women's Center, but she was not as comfortable in her work life. She left Polaroid and moved on to the Digital Equipment Corporation, a computer software company, where she did well and was promoted. Still, much of her job involved explaining new computer technology to much older men; and after five years in the field, she wanted to make a change. She missed doing science, and she was struck again by how few African American women were engaged in scientific disciplines. Almost ten years had passed since "The Double Bind," but diversity in science hadn't changed much.[4]

In the fall of 1984, Hammonds was invited to participate in a year-long seminar at the Center for Research on Women at Wellesley, a liberal-arts college for women located just outside of Boston.[5] In previous years the organizer, Peggy McIntosh, had brought together diverse groups of women to speak about other topics in women's studies. But despite her best efforts, this year's gathering of twenty-five participants included few women of color. In addition to Hammonds, they included Zala Chandler and Andree Nicola-McLaughlin, both from Medgar Evers College; and Celia Alvarez from the Center for Puerto Rican Studies at Hunter College.[6]

Evelyn Fox Keller attended several sessions, as did Helen Longino and Sandra Harding, both feminist philosophers of science. Though a number of the participants, including McIntosh and Harding, would later make significant contributions to scholarship on racism, white supremacy, feminist standpoint theory, and the concept of strong objectivity, they were at this time just beginning to think about these ideas within the context of gender and science.[7] Neither Ruth Hubbard nor Rita Arditti attended, perhaps because they were significantly older than many of the others, had been working on the topic of gender and science for more than a decade, or were merely busy with other things.[8]

The meeting room, with its large plate-glass windows, looked out over a lake—a calm setting, Hammonds thought. Nearby, a woman with close-cropped brown hair was sitting at a table with some other participants. Hammond thought she looked familiar. When she introduced herself as Anne Fausto-Sterling, a professor of biology at Brown University, Hammonds recognized her matter-of-fact voice, and her memory clicked into place: the two had met at the Boston Genes and Gender Group event in 1979, and Fausto-Sterling had been the person to call Hammonds about the seminar earlier in the year.[9]

The women at the seminar were grappling with the concept of objectivity in science. All were tugging on a thread dangling from what had once been a tightly woven fabric, a tapestry that Hubbard's writings had already started to unravel. Yet they were wary too. They were rethinking the entire intellectual structure of the sciences at a time when it was considered transgressive to question this ideology.[10]

Evelynn Hammonds certainly felt that she was not treated objectively in science. As a Black woman, she believed her comments were not taken seriously. As a feminist, she felt that her observations were seen as radical. Her own experience had shown that it was dangerous for a young woman in science to join the feminist critique, that she might jeopardize her ability to do the work that she wanted to do. Although Hammonds had not yet met Nancy Hopkins, she agreed that it was difficult for scientists to speak critically about their profession. If you were allowed into the club, then you'd better not be ungrateful and start criticizing.[11]

Hammonds felt that there had to be some way of supporting women scientists who were engaging in feminist critiques.[12] And, as it turned out, the Wellesley seminar became a useful discussion forum for those who were entering this difficult terrain. In small breakout groups, the women considered the implications of feminism on teaching and science curricula and how inclusive perspectives on race, sex and gender, class and sexuality might affect research and the teaching of science and technology. Each participant brought a different angle to the conversations. These discussions continued throughout the fall of 1984 and the spring of 1985. Some people would drop in for a session or two; others, including Hammonds and Fausto-Sterling, attended every one.[13]

The seminar drew top-notch presenters. The opening session featured Shirley Malcom, the co-author of "The Double Bind" report. Another featured

Sharon Traweek, an anthropologist studying physicists, who in 1988 would publish *Beamtimes and Lifetimes,* which famously described science as having "a culture of no culture." Sandra Harding and Helen Longino shared their work in progress, and the group discussed several of Evelyn Fox Keller's writings.[14]

In one session, Evelynn Hammonds led a discussion about women of color and science. In preparation, she asked all of the participants to read the "Combahee River Collective Statement" and Gloria Hull and Barbara Smith's article "The Politics of Black Women's Studies." Among other topics, she spoke of a Black woman scientist whom she had recently been reading about—the unusually named Roger Arliner Young, who in 1940 had become the first African American woman to achieve a PhD in zoology.[15]

This collective effort to question a discipline was taking place at a particular historical moment. In the early 1980s, the Cold War was intensifying. Civil rights and antiwar activism had ebbed during Ronald Reagan's presidency, yet American withdrawal from Vietnam was still fresh in everyone's minds, and the arms race had resulted in large protests against the use of nuclear weapons. *Resist authority* was a catchphrase, and the mainstream women's movement, now more than a decade old, was finally realizing that not all women were white and middle class. Black feminist writing was blossoming, thanks to Kitchen Table Press, which featured the voices of Alice Walker, Toni Morrison, Barbara Smith, and others. In 1981, bell hooks published her groundbreaking Black feminist manifesto, *Ain't I a Woman?* Women's studies programs were opening at universities and colleges across the country, and many feminists, including Barbara Smith, were promoting the need for Black women's studies programs. In 1984, Beverly Guy-Sheftall and Patricia Bell-Scott founded *Sage: A Scholarly Journal on Black Women.* At the same time, scholars were developing the interdisciplinary field of science studies. There was a growing interest in understanding the workings of science from philosophical, historical, and sociological perspectives. Science, technology, and society (STS) programs were opening at universities and colleges, and feminists were a part of this new wave of study.

The Wellesley seminar was an opportunity for women in science and women who were studying the philosophy of science to come together and ask similar questions. As they did, they built a network of relationships that would support them for years to come. Writing to Peggy McIntosh after the end of the seminar, Fausto-Sterling noted that her feminist colleagues at Brown were not scientists and that her scientist colleagues were skeptical of feminists: "The seminar provided a way for me to break out of the intellectual isolation I feel."[16]

At their annual conferences, various professional associations, including AAAS and the History of Science Society, would regularly invite a woman to speak on a panel about science and gender. Generally, that woman would then invite colleagues from various disciplines to join her. One year Ruth Hubbard was invited. In other years Sandra Harding, Evelynn Hammonds, and Anne Fausto-Sterling would have the opportunity, and they would invite members of their network to take part.[17]

Nonetheless, it was still possible for a group of white women to publish an anthology or sit on a panel at a conference about gender and not discuss issues of racial, ethnic, and class oppression. Although the Combahee statement had introduced the idea of intersecting identities, the theory of intersectionality would not take root until the 1990s. Audre Lorde's writings were beginning to have an influence; but at least in science, they hadn't been absorbed by a broad scholarly audience. Hammonds played an important role in transforming these attitudes, as did Sandra Harding. But even here, change arrived slowly. *Discovering Reality*, Harding's 1983 anthology on the philosophy of science, included Ruth Hubbard's "Have Only Men Evolved?" and Evelyn Fox Keller's "Gender and Science." Yet, overall, the essays displayed only a peripheral awareness of how race, class, ethnicity, sexuality, and culture intersect with gender to create knowledge and history. Harding later acknowledged the anthology's "Eurocentric heterosexist" perspective. By 1986, when she published *The Science Question in Feminism*, her awareness had shifted, and she was wrestling directly with issues of race, class, sexuality, and the other that had been missing from her earlier work. In Harding's case, the influence of postcolonial critiques in philosophy had been instrumental in shifting her thinking.[18] There was a growing awareness in academia that colonialism and imperialism had squelched other ways of being and thinking. But it would be years before this perspective had an impact on science studies.

For the women at the Wellesley seminar, these ideas were still in their infancy. But at least they now had someone to share them with. After rekindling their acquaintance, Hammonds and Fausto-Sterling sat down over lunch for a longer conversation. Hammonds talked about earning her master's degree in physics and her work as a computer programmer. The two pondered her interest in returning to science in some way, and the notion of

studying the history of science arose. Perhaps if Hammonds were to investigate the absence of African American women in science, she could make a
contribution. What if she were to earn a doctorate in the history of science?
Hammonds shared this idea with Shirley Malcom, who thought it was a
good one. Next Hammonds spoke to the physicist Walter Massey at Brown
University, who concurred. He introduced her to several historians, and she
decided to take the plunge.

Still, when Hammonds told Ruth Hubbard that she was applying to
Harvard, her response was "Why would you want to do that?" Hubbard
was expressing her feeling that Harvard was not a supportive environment
for women, but Hammonds thought Hubbard was saying that she couldn't
succeed there. The misunderstanding was ironed out when Hubbard clarified that she thought Hammonds was too good for Harvard. Nonetheless, a
warning hung in the air.

Hammonds decided to apply anyway and was accepted in the fall of 1985.[19]
She soon learned, however, that Hubbard wasn't the only person with mis-
givings. Some of Hammonds's friends discouraged her from attending Harvard; in their eyes, going there meant selling out. On this topic, Hubbard gave
Hammonds clear advice: it is important to do your political work wherever
you are. Hammonds found this very helpful. She wasn't going to stop talking
about gender and race just because she was at Harvard. And she didn't.[20]

As Hammonds prepared to begin her doctoral program, her mother
was diagnosed with ovarian cancer. Though she admired her mother's dig-
nity and courage, she also became despondent. During this period she was
buoyed by support from Audre Lorde, who, though twenty years older, had
become a friend. Diagnosed with breast cancer in 1978, Lorde published her
acclaimed *Cancer Journals* in 1980. When she came to Boston for speaking
engagements, Hammonds would pick her up at the airport and drive her
around Cambridge. The two would have long talks about coming to terms
with cancer, and Lorde encouraged Hammonds in her life and work.[21]

Despite her illness, Hammonds's mother gave her daughter her full support, confident that Evelynn would do well in her new career. Little did she
know that one of Hammonds's many achievements would be to bring an
awareness of race and racism not only to the history of science but to feminist science studies as well.[22]

FIGURE 1. Roger Arliner Young, Marine Biological Laboratory, Woods Hole, Massachusetts, circa late 1920s, courtesy Marine Biological Laboratory Woods Hole Oceanographic Institute.

FIGURE 2. Ruth Hubbard and George Wald, with the Paul Karrer Award, 1967, Courtesy Wald family.

FIGURE 3. Rita Arditti in the lab circa late 1960s, Courtesy Federico Muchnik.

FIGURE 4. Ruth Hubbard with George, Debbie and Elijah Wald at a Vietnam War protest, 1972, Courtesy Wald family.

FIGURE 5. Ruth Hubbard and student Kathy Kleeman in the lab, circa early 1970s, photography by Starr Ockenga, courtesy Schlesinger Library, Harvard University.

FIGURE 6. Rita Arditti, on the far left, with colleagues from New Words, circa 1974, Courtesy Federico Muchnik.

FIGURE 7. Dorothy Burnham in August, 1960, courtesy the Burnham family.

FIGURE 8: Lorraine Bethel, Beverly Smith, Demita Frazier, and Barbara Smith with Combahee River Collective Banner, Third World Women, We Cannot Live Without Our Lives, 1979, Photography by Ellen Shub, courtesy of the Estate of Ellen Shub.

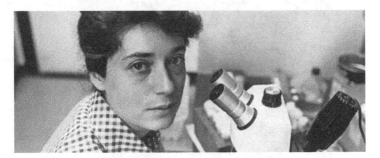

FIGURE 9. Anne Fausto-Sterling, circa early 1970s, courtesy Anne Fausto-Sterling.

FIGURE 10. Evelyn Fox Keller, 1990, Photography by Janet Van Ham.

FIGURE 11. Rita Arditti at the Massachusetts Breast Cancer Coalition march, September 12, 1993. Photography by Ellen Shub, courtesy of the estate of Ellen Shub.

FIGURE 12. Evelynn Hammonds at the podium, 1994. Courtesy MIT Museum.

FIGURE 13. Nancy Hopkins with fish tanks, circa 2000, Photography by Donna Coveney, courtesy MIT News and Nancy Hopkins.

FIGURE 14. Evelynn Hammonds at MIT conference, 2002, Photography by Laura Wulf, courtesy MIT News.

FIGURE 15. Evelynn Hammonds and Anne Fausto-Sterling at Fausto-Sterling's retirement symposium, May 2, 2014. Courtesy Anne Fausto-Sterling.

FIGURE 16. Banu Subramaniam, circa 2017. Courtesy Banu Subramaniam.

CHAPTER 13

READING, WRITING, AND
MYTH BUSTING

In the lab, the biologist Anne Fausto-Sterling's research focused on the development of fruit-fly embryos. But as she worked, she was also paying attention to the conversations around her. When the topic of feminism came up, her colleagues would sometimes cite studies concluding that male rats were more aggressive than females. She wondered if that difference really existed and if it were valid to use those studies to make conclusions about human behavior.[1]

Fausto-Sterling had completed her PhD in developmental biology at Brown University in 1970. Unlike Ruth Hubbard and many other older women, she did not get pushed onto a side track as a research associate. With the passage of the Civil Rights Act and the enforcement of Title IX, she was able to start her career on a regular academic track. At Brown, she taught biology and embryology and continued her research with *Drosophila*. And then came the Louise Lamphere case.

An assistant professor of anthropology, Lamphere was the only woman in her department at Brown, and in 1974 she was denied tenure. She filed a grievance; and after it was rejected, she brought a class-action suit against the university. At the time, Brown was routinely hiring women as assistant professors and then would not give them tenure. Only two women on the entire faculty were tenured. But in 1977, under the leadership of a new president, the university settled Lamphere's case before trial and agreed to a consent decree, which was designed to improve hiring and promotion practices and achieve full representation for women in the faculty at Brown. The decree remained in place for fifteen years, from 1977 through to 1992.

The Lamphere lawsuit was a critical turning point for Fausto-Sterling. In the middle of the case, she and five other women were promoted, and her tenure was confirmed. According to her, the suit gave Brown "a big kick in the butt."[2] By this time, she was becoming known in her field and had published in numerous scientific journals; yet, without Lamphere's case, she might have been left in limbo.

Fausto-Sterling's mother, Dorothy Sterling, had also wanted to become a scientist but had been discouraged from studying it in college because she was a woman. So she encouraged Anne and her older brother Peter (who became a neurophysiologist) to pursue both scientific research and social justice. Meanwhile, Dorothy Sterling undertook an alternate career, writing books for young people about nature, civil rights, and women's rights. In 1954, she published a biography of Harriet Tubman, and some of her books were banned at her own children's school.

During her sophomore year at the University of Wisconsin, Fausto-Sterling took a course on embryology and development. In class she watched a time-lapse film that transfixed her. It showed a single cell develop into an amphibian embryo. Amazed, she watched the cells replicate, grow, and build themselves into three-dimensional shapes that became the body of a frog. At that moment she decided this was what she wanted to study, and in 1965 she graduated with a degree in zoology.[3]

After earning her bachelor's degree, Fausto-Sterling began a master's program at Wisconsin but was uncertain about her thesis topic. Moreover, she was about to get married to Nelson Fausto, a medical pathologist. So when he was offered a job at a new medical school at Brown, she took a year off and moved with him to Providence.

As the wife of a faculty member, she had access to the university library, and she went there every day. One of the important books she read during this period was E. B. (not E. O.) Wilson's *The Cell in Development and Heredity*, which synthesized cell biology and developmental biology. The book had originally been published in 1925 (the year after Ruth Hubbard's birth), but in the 1960s it was still considered a classic..

Unlike some of his peers, Wilson did not head into genetics. He remained interested in the cell as a whole object and saw development as a holistic, interactive process. Fausto-Sterling was particularly taken with his chapters on maternal effects—that is, the effects of the egg and the cytoplasm on development. Based on her reading, she put together an idea for a thesis project that would look at specific fruit-fly mutations. She applied to Brown, was accepted, and began her work with *Drosophila*.[4]

Fausto-Sterling's career blossomed at Brown. But after she received tenure, many of her male colleagues recommended that she emphasize research rather than teaching. In their view, teaching was a burden, while work in the laboratory was a priority. Though she secretly enjoyed teaching, she took this advice.[5]

In a paper published in 1976, she described her lab project of breeding fruit flies with a unique mutation, one that caused a wing deformity. She was spending hundreds of hours staring into a microscope, meticulously measuring wing length and counting the number of hairs on each wing. She concluded that the fruit flies with this mutation had an overproduction of chitin, a material that makes the exoskeleton. She also noticed that nutrition had an unambiguous impact on the degree of wing distortion. In other words, even fruit flies with the genetic mutation were affected by the environment.[6] Although Fausto-Sterling did not mention nature and nurture in the paper, her conclusions illustrated the way in which both are entwined in the development of a fruit fly.

As her career advanced, Fausto-Sterling became more and more aware that nature and nurture were inseparable, and she worked steadily to shift both academic and public conversations about this topic.[7] Up until that point, however, she had kept her research and teaching separate from her personal work as a feminist and an antiwar activist. But after she received tenure, her priorities changed. Because she recognized that the classroom was a place to learn about the relationship between science and society, she began to emphasize her teaching. With her husband, she developed an extradepartmental course, open to all undergraduates, titled "Social Issues in Biology." The class dealt with issues such as occupational health, race and IQ, women and biological determinism, and sociobiology.

Inspired by this success, Fausto-Sterling then developed another new course, "Biology of Gender," which dealt with sexual dimorphism in nonhuman vertebrates as well as in humans, exploring concepts in genetics and evolution. But when she asked that the course be listed as a biology credit, she stirred up a hornet's nest among her colleagues, many of whom argued that her approach was not really scientific. Fausto-Sterling would not be swayed, and in the end she won the battle.[8]

In 1981, Brown established the Pembroke Center for Teaching and Research on Women and approved women's studies as an undergraduate concentration. Fausto-Sterling was the only woman scientist at Brown who was interested in participating in the new program. Though she had published numerous papers on embryology and development and would conduct research in the lab for the remainder of her career, she was also wondering how she could contribute as a scientist to this new area of scholarship. She was fortunate to have tenure by the time she began asking these questions: many scientists disliked bringing feminism into science, and working in women's studies was still seen as a liability for untenured faculty.[9]

Fausto-Sterling wanted to write a book that would explore how myths about biological sex differences had made their way into scientific thinking about biology. She knew she was well placed as a trained scientist to examine these preconceptions, but she didn't want to jump into a project that someone else might already be doing.[10] So she decided to consult with Ruth Hubbard. The two had met at the Boston Genes and Gender conference and as members of Science for the People, and she was familiar with Hubbard's writings, especially *Women Look at Biology Looking at Women*. She was also reading Ruth Bleier's feminist critiques on biology and Evelyn Fox Keller's essays on gender and science.[11]

Hubbard was very receptive to Fausto-Sterling's proposal. "We need your voice," she said. "We need as many voices as we can get." It was an important moment for Fausto-Sterling. Finally, she had found a way to combine her scientific skills with her work as a feminist.[12]

Other conversations were also influential—though not always comfortable. One day, for instance, Fausto-Sterling had lunch with a senior colleague in the Brown history department. He was interested in the new academic work in women's studies but couldn't figure out how it could relate to science. With sincere curiosity, he asked, "Don't you have to admit that, until Marie Curie, there had been no great women scientists? No women of genius?" She looked at him, stunned and uncharacteristically silenced. Every bone in her body felt that his assumption was wrong, but she had no way to prove it. Margaret Rossiter had not yet published her research, and there was little literature available, as both Evelynn Hammonds and Evelyn Fox Keller had also discovered.

Nonetheless, right after lunch, Fausto-Sterling started her own search. Every time she came across an article with a title such as "Five Great Women Mathematicians" or "Ancient Woman Chemist," she copied the piece and put it into a box. As the years passed, the pile of articles grew and grew, but her colleague's question still niggled. Then she came across an article titled "The Secret Life of Beatrix Potter" in the journal *Natural History*. In the late nineteenth century, well before Potter wrote *The Tale of Peter Rabbit*, she had been a natural scientist. In her research of lichens, she became the first person in England to observe and describe the symbiotic relationship between algae and fungus, traveling throughout the British Isles to classify fungi and making detailed drawings to identify types. Then, after she submitted her groundbreaking paper on lichens to the Linnaean Society of London, she received word that she could not share her findings there because women were not permitted at the society's meetings. In the end, her uncle read the paper. But Potter, frustrated by the barriers to women, including lack of access to the British Museum, abandoned science and turned to writing children's books.

The story flicked a switch in Fausto-Sterling's head. The Potter story was a tale of erasure. Who else had become invisible? How many other women's work had been lost from view? Never again would Fausto-Sterling be silenced by a question about the lack of women in science. Her epiphany led to her growing interest in the history of science—a fascination she shared with Evelynn Hammonds during their conversations at Wellesley.[13]

In the early 1980s, two new books had a huge impact on many women scientists, including Fausto-Sterling and Hammonds. The first was Rossiter's *Women Scientists in America: Struggles and Strategies to 1940* (1982). It "blew me out of the water," recalled Fausto-Sterling. "I couldn't put it down. It confirmed for me how wrong that history colleague of mine was. There were women everywhere and sometimes they were hidden in the laboratory budget as equipment."[14] The book offered a detailed chronicle of the history of white women scientists in the United States from 1820 through to 1940. In it, Rossiter argued that there was a clear link between seeing women as subordinate in science and their historic invisibility, even among experienced historians of science. In the nineteenth century, she said, "[a] camouflage [was] intentionally placed over their presence."[15]

Until reading Rossiter's book, Hubbard, like Fausto-Sterling, had always been more interested in how science perceived women's biology. She hadn't been particularly intrigued by women scientists or women in science. But *Women Scientists in America* changed her mind. She was fascinated by what Rossiter had documented. Since 1820, women had been making steady, careful, and politically astute efforts to join the scientific world, but barriers had been thrown up at every turn. Long before Hubbard had been cut out from her husband's Nobel Prize, women had been struggling to be included and recognized in the world of science. She began to reflect in new ways about that history and her own generation's similar challenges.[16]

Then, in 1983, Kenneth Manning published *Black Apollo of Science: The Life of Ernest Everett Just*. Manning had been the first Black person to earn a PhD in the history of science from Harvard, and the subject of his research, Ernest Everett Just, had been a Black biologist from the 1920s through the 1940s, conducting cell research at Woods Hole and teaching at Howard University. Both Fausto-Sterling and Hammonds read the book, and they were impressed by Manning's research and Just's life story. At the same time, however, they were both taken aback by how Manning portrayed the young scientist Roger Arliner Young, the first Black woman to earn a PhD in zoology. Though she, too, had taught biology at Howard and conducted research at Woods Hole, Manning offered very little information about her life. The biographer's defense, according to Fausto-Sterling, was that the book was not about her.[17]

In 1985, when Evelynn Hammonds started her PhD in the history of science, she was particularly interested in the history of Black women scientists. She'd read Rossiter's book and noticed, of course, that it didn't include any Black women or mention issues of race. Hammonds had already done some digging and had learned that five Black American women had earned doctorates in science before 1940—among them, Roger Arliner Young, who'd had little support or encouragement during her years at the University of Chicago. Hammonds also learned that many Black women without the opportunity to pursue PhDs had nonetheless worked in laboratories and scientific institutions. She was interested in which institutions Rossiter had focused on during her research and how she had written about the barriers to entry for women. She wanted to show that the barriers were very different for Black women, most of whom had to pursue their studies in Black colleges without the resources for laboratory research.[18]

In *The Black Apollo of Science*, Manning had revealed how white male scientific leaders in the early twentieth century had decided that Black people in historically Black colleges should not become researchers but teachers. At the time, large foundations were funding research infrastructure in a range of universities across the country, but they chose not to do so at Spelman, Morehouse, Howard, and other historically Black institutions. In the early 1920s, Ernest Just was seriously considered for a position at the Rockefeller Institute, which would have allowed him to do dedicated research for several years, but he was turned down and encouraged to focus on teaching instead. These funding decisions had sweeping implications, preventing the full participation of African Americans in science. This was exactly the kind of issue that Hammonds wanted to explore among Black women scientists. But in the mid-1980s, most historians of science, even the feminists, weren't particularly interested.[19]

Rossiter's *Women Scientists in America* had made the case that the development of women's colleges in nineteenth-century America had given women an "entering wedge" into the increasingly professionalized world of science. By the end of the century, she wrote, there were enough science professorships at the major women's colleges to form the basis of a small network of female scientists. But Hammonds knew that this trend did not hold true for Black women's colleges. In southern women's institutions in the 1930s, expectations, resources, and opportunities were very different from those at New England women's colleges. Science was not offered as an option.[20]

Now that Hammonds had learned about Roger Arliner Young, she tried to gather all the information she could, and Manning helped. She discovered that a falling out with Just had ended Young's tenure at Howard and Woods Hole. After that, her opportunities in science had been limited, and Hammonds wanted to find out more about what had happened.

Meanwhile, at Brown, Fausto-Sterling was developing a course titled "Women and Minorities in Science." She had long wanted to teach such a class but had been stymied by the lack of good reading material. Now, however, with the Rossiter and Manning books out, she had reconsidered.[21] Fausto-Sterling wanted the course to serve as a practical bridge between women's studies—which all too often meant *white* women's studies—and what was then called Afro-American studies. So she worked with one of her students, Lydia English, to develop a reading list and a syllabus. English was an African American woman who had worked in banking for ten years before returning to Brown to study anthropology.

The course drew eleven students in the spring of 1985: three Black men, five Black women, one Asian American woman, and two white women. Most were seniors majoring in biology, engineering, computer science, or biochemistry. Rather than being based around lectures, the course required students to share presentations, summaries, and analyses of the readings with the group. The first section considered biographies and profiles of Black and/ or female scientists, and Fausto-Sterling and English were well aware that the written material about Black women scientists was weak. Moreover, Manning had portrayed the women in his book in ways that made them both uncomfortable, and the issue of race was "glaringly absent" from Rossiter's book.[22] To add needed context, Fausto-Sterling invited Hammonds to speak to the students, and Hammonds agreed. In her talk, she compared her undergraduate experience at Spelman, a historically Black women's college, with her graduate experience at MIT, a historically white institution. She also spoke about her interest in Roger Arliner Young.

Hammonds and Fausto-Sterling had already developed a friendship via their conversations about feminism and science. But in 1984, after Anne's mother, Dorothy Sterling, published *We Are Your Sisters*, an anthology of writings about nineteenth-century Black women, they found that they could also discuss race and racism, topics that Hammonds couldn't talk about easily with some of her other feminist colleagues. The friendship became an important intellectual and emotional outlet for them both.[23] One day Fausto-Sterling shared a *Scientific American* article with Hammonds about sexual behavior among a species of lizards. Even though the lizards were a unisexual species, the author saw them through the lens of human heterosexual relationships. Both women recognized that the author's views were totally biased, but they also found them funny. Even while they understood the potential damage such articles could cause, it was good to be able to laugh together.[24]

In 1985, as Hammonds was beginning her PhD, Fausto-Sterling published *Myths of Gender*, the book that Hubbard had encouraged her to write. In it, she carefully examined current research about sex differences relating to biology, hormones, and genetics and concluded that there was a shocking lack of substance behind ideas about biologically based sex differences. In the introduction, Fausto-Sterling recalled that some nineteenth-century scientists believed that "women who work to obtain economic independence set themselves up for a 'struggle against Nature,'" and she pointed out that many people had used Darwin's theory of evolution to argue that giving the vote to women was, "evolutionarily speaking, retrogressive." Her book rejected the nature-nurture framework, arguing instead for a more complex analysis of the way in which individual capacities grow from a web of interactions between biology and social environment."[25]

Myths of Gender addressed the many cases of biased scientific research about male-female biology that had been picked up by the popular media. For instance, in her book's review of research on sex differences in math ability, Fausto-Sterling cited a 1980 *New York Times* article claiming that boys were genetically better at math than girls were. She recalled that, when the article was first published, a friend had asked her, "Is it true? What can I tell my daughter?" After taking a closer look at the research behind the claim, Fausto-Sterling pointed to many problems with the methodology and conclusions.[26] She also discussed "The Sexes: How They Differ and Why?," a 1981 *Newsweek* cover story. As it happened, the article itself was much less conclusive than the headline. The reporters had written, "Perhaps the most arresting implication of the research up to now is not that there are undeniable differences between males and females but that their differences are so small relative to the possibilities open to them."[27]

In the book, Fausto-Sterling systematically examined research on hormones and psychology, hormones and menopause, and hormones and aggression. She then made three important points. First, scientists are influenced by the cultural lens through which they conduct their research. Second, any observed differences between men and women are not clear and consistent. And, third, any biological differences are shaped and influenced by the social environment and therefore vary greatly across men and women.[28] She also devoted an entire chapter to sociobiology, picking up on some of the arguments that Ruth Hubbard, Rita Arditti, Marian Lowe, and Freda Salzman had made after E. O. Wilson's 1975 publication of *Sociobiology* and taking great exception to Wilson's 1978 follow-up *On Human Nature.*

Hammonds applauded the book. She felt that the style of *Myths of Gender* had the potential to reach a broad group of women who were not scientists, helping them understand and talk about the topic. In this way, they, too, could shape public opinion and create a more popular feminist critique of science.[29]

In January 1987, Evelynn Hammonds spoke at a conference on "Women in Science" held at Hampshire College in Amherst, Massachusetts, a small progressive institution that had become a node of activity related to women and science. Mary Sue Henifin, Hubbard's co-editor on *Women Look at Biology Looking at Women* and *Biological Woman: The Convenient Myth*, was now an assistant professor in its biology department and, along with her colleague, Nancy Goddard, had developed a course called "The Biology of Women." Throughout the 1980s, she hosted a series of "Women in Science" conferences, which took place every six months at several colleges throughout New England.[30]

The January conference focused on the topic of race, gender, and science, and Hammonds shared some of her early research on Black women scientists. Fausto-Sterling and Rita Arditti were in attendance and gave her good feedback on the paper. Other participants included Vanessa Gamble, a medical doctor who was about to finish her PhD in the history of science, and Darlene Clark Hine, who had built the field of Black women's history. Hine later asked Hammonds to write an entry on Roger Arliner Young for her encyclopedia of *Black Women in America*.[31]

But there was also some uneasiness at the gathering. The conference organizers—Ann Woodhull-McNeal, a professor of biology, and E. Frances White, a professor of history and Black studies, had drafted a review of *Myths of Gender* for the magazine *Radical America* in which they chastised Fausto-Sterling for not including race in her discussion of gender. She had seen the draft and was upset by the criticism; she felt that it suggested that she should have written a different book.[32] The editor of *Radical America* had asked Arditti and Hammonds to read the draft review. Hammonds thought the tone was unnecessarily strident and that their examples from other writers implied that Fausto-Sterling had "gotten it all wrong." Arditti also thought that certain sections of the review were too harsh. So, according to Hammonds, when they all arrived at the conference, there was serious tension in the air.[33]

White defended the tone of the review, saying that it reflected the feeling that she "couldn't see herself" in the text, but she did agree to revise the review for publication. In its final form, it began with high compliments for the book's description of how scientific theory and observation are influenced by social factors. The reviewers praised Fausto-Sterling for challenging the myths that white western society had promoted about the "nature of women."

They applauded her for systematically critiquing the so-called scientific evidence that was often used to bolster these myths. They did, however, raise concerns about the absence of any consideration of race and suggested that Fausto-Sterling had written as if a single set of myths about gender applied to all women. The reviewers were concerned that science had played a role in constructing distinctly negative images of Black women, and they wished that her book had shed light on that history.[34] White and Woodhull-McNeal pointed out that *Myths of Gender* addressed numerous myths about women as weak, dependent, emotional, and nonmathematical. But the myths about the nature of Black women were very different, and feminist scientists had not given the issue much attention, even though Dorothy Burnham had pointed this out at the Genes and Gender Collective conference ten years earlier.[35]

Since Audre Lorde's 1979 talk at "The Second Sex" conference, such concerns had slowly—very slowly—started to become part of the discussion among women in science. Hammonds was also grateful for her personal friendship with Lorde, who had encouraged her to challenge the silencing of Black women in the women's movement. Their conversations had given her strength when she was working among white male scientists and white women historians of science.[36] Lorde had also given Hammonds the courage to address sexism within the Black community, especially in the workplace. She had helped Hammonds see that, for Black women, sexism and racism were not separate issues.

In 1988, Hammonds spoke about Lorde on a radio show. In the course of that conversation, she mentioned a moment in the workplace when a group of white women had spoken up about sexual harassment. When a Black male colleague advised Hammonds and other Black women who had been harassed not to create an alliance with the white women, she interpreted his comments to mean that Black women's experiences should not "see the light of day" because they were not a priority for Black men. This Black man, she believed, was ready to accuse her of being disloyal to the Black community. At the time, this realization had felt like a crisis, as if she really did have to choose between standing up for the Black community and standing up for the women's community. But with Lorde's influence, she came to understand that she didn't have to agree with this man's opinion of what it meant to be a loyal member of the Black community. She decided to align herself with the Black women who were being harassed; and if that meant working with white women, then too bad. Of course, in many cases, such decisions weren't always so straightforward.[37]

Soon after Hammonds started working on her PhD, she took a course on gender and science taught by Evelyn Fox Keller at MIT. Hammonds had read Keller's essay "The Anomaly of a Woman in Physics" and was excited to spend time with her. She knew that both of them were feminists and that both had studied physics. And she was right to be excited: Keller's class turned out to be one of the most demanding and intellectually stimulating courses she had ever taken.[38]

After the disagreement at the Hampshire conference, Hammonds wrote to Keller, suggesting that the two of them meet with Fausto-Sterling to discuss issues of race, gender, and science. But that meeting never happened. In the end, despite their commonalities, Hammonds and Keller were never able to talk about race and racism with the ease that Hammonds shared with Fausto-Sterling.[39] Nonetheless, Hammonds's own research and publications were not only echoing some of Keller's experiences in physics but breaking new ground as well.

In 1986, the journalist Aimee Sands interviewed Hammonds about her experience as a Black woman scientist. Hammonds not only talked about her life in physics but also examined the challenges facing Black women scientists, as distinct from those facing white women, an area of analysis that previously had received scant attention.[40] Sands's interview, "Never Meant to Survive: A Black Woman's Journey" (the title quoted from a Lorde poem), appeared in numerous collections and anthologies.[41] Like Keller's biographical essay, it became one of the foundational pieces of feminist science studies.

During these years, Hammonds also got to know Rita Arditti. Arditti owned a triple-decker apartment near Porter Square in Cambridge, and she lived on the third floor. Generally, this style of building has a single front door that leads both to the first-floor apartment and to the stairway to the upper floors. This makes it easy for all of the tenants to get to know one another. After Hammonds moved into the second floor of Arditti's house, she learned that the people on the first floor often hosted Science for the People meetings.[42] The house was full of science talk.

More than fifteen years had passed since Arditti had led the protest at the AAAS conference. Hammonds, who was just getting to know her, found her to be kind and warm but also intense. In their conversations, Arditti was hopeful that science, over time, would become more egalitarian. Hammonds wasn't sure that this was possible with men at the helm.[43]

In the 1980s, Arditti was still working at New Words bookstore and still living with breast cancer, alternating between periods of illness and relatively good health. In their activism around cancer, she and Lorde had also recently crossed paths. Arditti was no longer engaged in lab research, but she did continue to teach science at the Union Graduate School. Meanwhile, her son, Federico, now grown up, was producing videos for public television and performing as a singer-songwriter around town.

In 1980, when the historian of science Carolyn Merchant published *The Death of Nature*, Arditti read it immediately. Several historians and feminist scholars have since dated the origins of feminist science studies to this book, and Arditti, too, was affected by it.[44] In a review published in *Science for the People*, she applauded Merchant for explaining how the scientific revolution, capitalism, and industrialization had affected society's worldview—for instance, in encouraging the exploitation of Earth's natural resources. Merchant linked her historical work to the current ecological crisis, calling for a more holistic view of humans and nature, and Arditti welcomed her call for a return to a more interconnected view of the world.[45]

The year 1980 was a momentous one for Arditti. At a cancer workshop, she met Estelle Disch, a sociologist at the University of Massachusetts Boston, and the two eventually became life partners. In the same year, Arditti published the co-edited anthology *Science and Liberation*, which Fausto-Sterling used in her own teaching. A quote from Arditti's essay summarized the message of the entire book:

> The task that seems of primary importance—for women and men—is to convert science from what it is today, a social institution with a conservative function and a defensive stand, into a liberating and healthy activity.... When science fulfils its potential and becomes a tool for human liberation, we will not have to worry about women "fitting" into it because we will probably be at the forefront of that "new" science.[46]

By the time Hammonds began her PhD program at Harvard, she was in her early thirties and had come out as a lesbian. Her mother was perplexed by this announcement. Evelynn was already a Black woman in a nontraditional profession. Why would she want to make her life even more difficult, not to mention alienate herself from the Black community? Fortunately, Hammonds had support from both Audre Lorde and Beverly Smith.[47]

In October 1987, Hammonds joined hundreds of thousands of people in Washington, D.C., to march for lesbian and gay rights. During this event the enormous AIDS Memorial Quilt was displayed for the first time. Laid across the National Mall, its 1,920 colorful fabric panels were stitched with the names of those who had died of AIDS. Hammonds felt the sadness that the quilt represented but also loved its beauty and its affirmation of gay life.[48]

Back in Cambridge, Ruth Hubbard had asked Hammonds to serve as a teaching assistant for her class "Biology and Women's Issues," and Hammonds had agreed. Watching Hubbard lecture, she reflected on what an incredible teacher she was. With her long hair in a braid, and while clearly not wearing a bra, she spoke of the nineteenth-century women who deformed their bodies with corsets. Hammonds wanted to laugh: "Her lectures were great, and she enjoyed her students. Ruth was wonderful in the classroom and I learned a tremendous amount from her."[49]

In her own presentation to the class, Hammonds began by saying, "This is the first time I'm giving a lecture here at Harvard so bear with me." To her surprise, Hubbard was furious with her after the session was over. "Don't ever do that again," she said. "You get up there and do what you need to do. Do not apologize for anything." From then on, recalled Hammonds, "I worked my ass off for her because I loved it so much."[50]

Although the two women shared common interests, Hubbard, as a member of the biology department, couldn't serve as Hammonds's doctoral supervisor in the history of science department. And she was feeling pessimistic about her own situation. In the fifteen years since she had received tenure, things had gotten worse for her women colleagues. An outsider in her own department, she had little impact on decision making. Hubbard continued to sound off like a feminist broken record as her male colleagues continued to define "best person for the job" in such a way that every new appointment was a man. Harvard would only change, she thought, when the rest of society changed. So she focused her energies elsewhere, starting a new organization called the Council for Responsible Genetics.[51]

Meanwhile, Hammonds had her own departmental struggles. She'd received encouraging feedback on the paper she'd delivered at the Hampshire College conference, and now she wanted to spend all of her time researching the history of Black women scientists. However, her supervisors and her peers suggested that this would not be a good topic for her PhD. Hammonds later said that every graduate student at the time who was interested in writing a dissertation on women or people of color in science was urged against it.[52] Although Ken Manning's book about Ernest Everett Just had been a finalist for the Pulitzer Prize, other scholars did not see this as an opportunity to open up a new area of research into Black scientists. Even with Rossiter leading the way, the study of the history of women in science was still in its infancy.[53] Hammonds was frustrated and discouraged. Her supervisors weren't saying that these topics were unimportant; rather, they were saying that they would not be received as important in the broader field of the history of science and therefore not solid topics on which to build a career.[54]

Once again, Hammonds was battling to succeed in her chosen field. She had gone from physics where there were very few people of color, to computer science, where she found a similar situation, to the history of science, where she was the first Black woman to pursue a PhD in the department. And now she wasn't getting any support for studying gender and race.

Feeling trapped and needing to "figure a way out" Hammonds focused on finding a new direction, and she began by examining what other questions in the history of science were important to her. She'd always been curious about how scientific knowledge was translated into medical practice, public health, and public understanding. Now her supervisor, Barbara Rosenkrantz, suggested that she look into the history of diphtheria, one of the first diseases to be controlled as a result of new scientific knowledge in bacteriology. Perhaps Hammonds could look closely at how these discoveries had influenced public health in New York City. Rosenkrantz herself had worked a little on the diphtheria question, and she shared two boxes of documents. As Hammonds immersed herself in her new research, she became more and more interested.[55]

Meanwhile, in the late 1980s, the HIV/AIDS epidemic was having a major impact in the United States, and Hammonds started writing and publishing extensively on this topic. She discussed race, sex, and the construction of other in relation to popular perceptions of how the virus was spread. She considered the ways in which the mainstream media had failed to challenge the "age-old American myth of blacks as carriers of disease." She suggested that Black media outlets had been slow to take up a discussion of AIDS in the Black community because they were concerned about a backlash—that Black people would again be associated with "disease and immorality."[56]

In a course on the history of epidemics, Hammonds presented a paper showing that African American women were thirteen times more likely than white women to die of AIDS. The response from the other students stunned her. What she was saying couldn't possibly be true, they argued. The students argued that the AIDS epidemic was about white gay men and that few Black women were dying from the virus. The underlying message was that her paper wasn't important.[57]

Although the focus of her PhD topic had shifted to the history of diphtheria, Hammonds continued to apply her feminist critiques to this new area of work. She taught a graduate seminar using the writings of Hubbard, Fausto-Sterling, Keller, Rossiter, and Sandra Harding. Her interest in the history of Black feminism and the knowledge she'd acquired from the Combahee River Collective and Audre Lorde also informed the way in which she looked at the history of science, medicine, and public health. Though she was now out of the lab herself, she could still relate to the women scientists who were trying to find their way there. She recognized the conflicts between women scientists and feminist critiques of science. She saw there were tensions among various feminist critiques as well.[58]

CHAPTER 14

TENSIONS IN FEMINISM

The late 1980s was a period of uncertainty for Evelyn Fox Keller. Though graduate students such as Evelynn Hammonds truly valued the classes she was teaching at MIT, Keller's position as an adjunct lecturer was uncertain. Then, in April 1987, she received a letter from the Office of the President stating, "I am pleased to inform you that you have been reappointed Visiting Scholar, part-time, for the period February 1, 1987 to June 30, 1988, without salary." Keller needed a salary; and in her search for full-time employment, she reached out to several colleagues, among them Anne Fausto-Sterling. Ironically, just a year earlier, Brown University had reached out to Keller, stating that Fausto-Sterling was being considered for a promotion and asking for her opinion. In a generous response, Keller called Fausto-Sterling "one of the pioneers" in the new field of gender and science, emphasizing that her contributions to that field were unique, given that she was also a practicing research scientist in developmental biology. Fausto-Sterling's enquiry in gender and science, Keller wrote, was grounded in "familiarity with and respect for scientific procedures." Thanks to this kind of support, Fausto-Sterling did become a full professor at Brown.[1]

Meanwhile, Keller would search for an institutional base for another five years, leaving MIT for stints at Princeton and the University of California at Berkeley. Ruth Hubbard wrote, "I'll miss you a lot even though we didn't see that much of each other," and sent her articles to read and review. Keller assured Ruth she would miss her too: "The endurance of our friendship, and our ability to talk through (or around) disagreement, remains a kind of model for me."[2]

The two shared many areas of interest, yet they never seemed to be entirely on the same page. Hubbard was still focusing on the science of gender, Keller on the gender of science. In their discussions, which ranged from women's bodies to genetic engineering, there were always points of disagreement. "We've never really hashed it out," wrote Keller. "Perhaps we could do better in letters." Via mail, they continued their edgy arguments. "I was troubled and amazed by your saying that for you bombs are part of nature," wrote Hubbard. "I look on bombs as cultural artifacts like washing machines or tennis balls. Of course, all of them are parts of nature in the trivial sense that they are human creations and we are part of nature. But to me being killed by a bomb is . . . not like being struck by lightning." In another letter, sent months later, she asked, "Are we miles apart or saying similar things differently?"[3]

In 1988, during a drive from Cambridge to Rhode Island, the two had what Hubbard later called an absurd conversation about sex, gender, and feminist theory. Keller was committed to feminist theory in the academy and felt that Hubbard had no connection with it.[4] In a follow-up letter, Hubbard retorted, "If feminist theorists stop being able to converse with ordinary feminists and ordinary women, then we are in trouble. . . . I believe that the purpose of feminist theory is to strengthen women's liberation, not to build yet another academic specialty." Previously, despite their disagreements, she had always signed her letters "Love, Ruth," but after that road trip, she simply signed "All the best."[5]

Like Keller, Hubbard continued to think of herself as a scientist. Nonetheless, she did not supervise any graduate students, perhaps because she was a member of the biology department whereas most of the students who were interested in her work came from other disciplines. Had she been able to take on graduate students, there would have been a growing number of people studying science from a feminist perspective. But her legacy was elsewhere. Throughout the 1980s, she continued to publish papers and articles on aspects of biology and women's issues, reproductive rights, and genetics. Along with Rita Arditti, Jon Beckwith, Richard Lewontin, and others, she served on the editorial advisory board of *Science for the People* until both magazine and organization shuttered in 1989. And she compiled a collection of her essays, which were eventually published in 1990 as *The Politics of Women's Biology*. She gave frequent talks and presentations, and she continued to influence students. Sandra Harding later said that Hubbard was one of the first science studies scholars to take up the issue of race.[6]

But she was getting ready to retire; and in June 1990, her friends, students, and colleagues organized a day-long retirement symposium for her at the Radcliffe Quad. The attorney Margaret Burnham was among the speakers. "Ruth," she said, "has drawn with tremendous clarity the link between racism and chauvinism against women in scientific thinking." Hubbard's practice was "to close the artificial gaps between those of us who place ourselves in the women's movement and those of us who place ourselves in the anti-racist movement. Ruth knows that those lines are artificial and eminently permeable." Burnham noted that, when Mel King became the first Black candidate to run for mayor in Boston in the early 1980s, promoting support for low-income communities, Hubbard had been an ardent supporter of his campaign.[7]

"What I want to say about Ruth is that she is not Ruth but Ruthless," said Richard Lewontin to gales of laughter. "She is a ruthless radical, and for me that is the highest possible praise." Lewontin applauded Hubbard's "ruthless radicalism," not only because she had reexamined every assumption but also because she was radical in her teaching, her scholarship, and her activism. He said that she was the only person on the Harvard faculty to whom he would turn for political advice.[8]

After retiring from Harvard, Hubbard continued her writing, speaking, and activism, especially around genetics. She published several more books, including *The Politics of Women's Biology* (1990) and *Exploding the Gene Myth* (1993), which she co-wrote with her son, Elijah Wald, and *Profitable Promises: Essays on Women, Science, and Health* (1995). Her network of feminist students and colleagues fanned out at universities across the country, yet many scientists remained unaware of her work.[9]

As Hubbard was retiring, Evelynn Hammonds took part in a celebration of Audre Lorde's life and work. With Barbara and Beverly Smith, Rita Arditti, Estelle Disch, and many others, she helped to organize a four-day "I Am Your Sister" conference. More than a thousand people, including Lorde, squeezed into an auditorium in downtown Boston for the opening festivities. Then the work sessions shifted to the MIT campus, and there more than four hundred people crammed into a seminar room for a presentation titled "Feminist, Antiracist Organizing."[10]

In the same year, Hammonds partnered with Helen Longino, a philosopher of science, to co-write the seminal essay "Conflict and Tensions in the Feminist Study of Gender and Science." The two had met at the Wellesley seminar, and their joint effort appeared in the anthology *Conflicts in Feminism*, edited by Evelyn Fox Keller and Marianne Hirsch—an important reference work in the field of feminist science studies.

According to Longino and Hammonds, feminist thinking about science had made great contributions during the past fifteen years, producing many new ideas and analyses of how women had been subordinated in the natural sciences. Yet despite gains in some areas (notably, biology), the sciences continued to be "bastions of masculinity." Women made up only 20 percent of the biologists in the country, 4 percent of physicists, 8 percent of chemists, and 8 percent of engineers. Longino and Hammonds argued that the content of science remained hostile to women—from biological research arguing that gender inequality was the result of genetic and physiological differences to the metaphor that equates science with "male sexual conquest." They also noted that conflicts among feminist scholars in the field could be sharp and that at least one scholar had left the field to "get out of the crossfire."[11]

In one section of the essay, Longino wrote about internal feminist debates, asking questions such as, What is the goal of feminist critiques of science? Is it to create social change? Or increase access for women? Or produce science that does not "diminish, neglect or pathologize" women? Or produce an alternative science with different priorities? She described the work of four feminists with different viewpoints: Anne Fausto-Sterling, Evelyn Fox Keller, the biologist and science studies scholar Donna Haraway, and Sandra Harding. She concluded that all four believed that the sciences had been shaped by the concerns and interests of European and American middle-class men. All claimed that the sciences were part of the "structures of power" that feminists wished to change, but they disagreed on the best way to do this. Longino referred to Audre Lorde's declaration that "one cannot use the master's tools to tear down the master's house" and suggested that science was still the master. In Longino's telling, Fausto-Sterling and Keller, both committed scientists, were looking to reform science from within, whereas Haraway and Harding were more interested in shaking the foundations from outside. Given that the field of "feminist science studies" was still very small, she hoped that these disagreements could be used to strengthen the discipline rather than "[foster] competition over limited concepts."[12]

In her section of the essay, Hammonds focused on issues facing women in science that increased the likelihood that they would not engage with any feminist critiques. The women in science were likely to ask, What is it about women and women's lives that have kept them from doing science? In contrast, feminist critics of science were likely to ask, What is it about science that has limited the participation of women and, by extension, other marginalized groups?[13] Many women scientists—among them, Nancy Hopkins—believed that such feminist critiques reinforced traditional stereotypes about women and science, hurting rather than advancing their cause.

Hammonds pointed out that many working women scientists continued to see no connection between the problems that women encountered and "the structure of scientific knowledge and scientific education." They persisted in arguing against the perception that women can't do science while never addressing the source of such perceptions. Their comments suggested that "the perception of exclusion lies with the excluded," not within the profession.[14]

Hammonds used a review in *U.S. Woman Engineer* of Anne Sayre's 1975 biography of Rosalind Franklin to illustrate that many professional women scientists didn't know of or understand the feminist critique. Their deep belief in a meritocracy in science was, she said, "unexamined and unshaken." The reviewer concluded that "today we can look at Franklin's story as a lesson to keep an open mind, to listen to others, to avoid being overly dogmatic; and most importantly to speak out when one is sure of the facts—not an easy balancing act!" Yet as Hammonds noted, this writer clearly did not know the complicated circumstances that had "led to the suppression of Franklin's contribution to the discovery of the structure of DNA." Instead, the reviewer reduced Franklin's story to a "lesson in assertiveness training" for women who want to be successful scientists.[15]

According to Hammonds, professional journals for women scientists and engineers were continuing to reinforce the notion that women could do science well even while performing all of their other so-called womanly duties. These journals didn't question the structure of scientific institutions or examine whether or not they were supportive of women. Some portrayed feminist critics as accepters of stereotypical ideas about women rather than questioners. "This is a very curious reading of the feminist literature," pondered Hammonds.[16]

She also noted that, because many examples used in feminist critiques of science came from biology, women scientists in physics and mathematics found it easy to reject them. "For these women scientists, science is defined as the product of scientific method—a method that self-corrects for all human biases—including any that could arise from biases of gender." Hammonds shared the story of a mathematician who rejected any possibility of a useful feminist perspective by saying, "One still wants to know whether feminists' airplanes would stay airborne for feminist engineers." In short, she was arguing that feminist critiques had a larger purpose than simply making it possible for more women to do science.[17]

Hammonds and Longino concluded that most scientists, both women and men, had been educated without any reflection on the history of their own discipline. In this, they agreed with Margaret Rossiter, who had suggested that all of the work that women scientists had done in the early twentieth century to support women in the sciences and to lobby scientific organizations to recognize women had failed. Science was not objective or neutral, and it reflected social, economic, and political interests. "What options do practicing women scientists have," asked Longino and Hammonds, "other than to apply the existing method more faithfully?" Feminist critics had done well in articulating that gender was part of scientific practice, but they had done less well at demonstrating that the scientific method itself was marked by gender.[18]

The volume editor, Evelyn Fox Keller, agreed with this analysis. In her contribution to the collection, she wrote that, when she'd first started writing about gender and science fifteen years earlier, she'd believed that her primary audience would be other women scientists. She'd felt that if she could not reach them, her efforts would "constitute a fundamental failure." Now, however, Keller believed that this goal was almost impossible. Feminists who had begun as scientists (such as Keller, Hammonds, Hubbard, and Arditti) increasingly sensed that they were being forced to choose between doing work on feminist theory and doing science.[19]

Donna Haraway was another scientist whose feminist critique had taken her away from her work in the lab. Decades later, Haraway wished she had "taken more care" not to come across as anti-science. She wished she'd made scientists feel more welcome in the conversation.[20]

By the end of the 1980s, a growing movement of feminists involved in science had begun to shape teaching in science studies and women's studies. Yet while the writings of scientists such as Hubbard, Fausto-Sterling, Keller, and

Hammonds and the commentary of historians and philosophers of science such as Rossiter, Harding, Haraway, and Longino were regularly included in such reading lists, they were not yet rocking the foundations of science. The master's house was still firmly in place. Moreover, the feminist movement in science still had limited awareness of the impact of racism and discrimination within the sciences. Nor was it engaging sufficiently with international developments and trends or extending its critiques beyond biology and medicine into chemistry, physics, and engineering.

Despite these concerns, Evelynn Hammonds was excited to begin a theoretical critique that she hoped would broaden to include all of the sciences. She thought that the field of feminist science studies was being born, that an academic leadership would form, that there would be an explosion of scholarship as more and more people took part. Ideas were spreading, and a younger generation was joining the charge.[21]

PART THREE

FEMINIST SCIENCE STUDIES COMING OF AGE, 1986 TO 2005

CHAPTER 15

DEVELOPING NEW TOOLS

On a warm August night in 1986, Banu Subramaniam boarded a plane at the Bombay Sahar International Airport. She was flying across the ocean to America, where she would attend graduate school at Duke University. As a child, she'd watched the awe of David Attenborough and Jacques Cousteau as they presented the natural world to television viewers, and she learned that science and technology had been important for India's growth and development. In high school, she'd read *On the Origin of Species* and had decided then and there she wanted to study evolutionary biology. While at the University of Madras, she'd attended an exhibition about Darwin at the U.S. consulate and they had given her a beautiful poster of him. For years, she'd sat at her desk, working under that giant poster of Darwin, her role model and idol.[1]

Back then, Subramaniam was enraptured by the world of science, its rationality and objectivity. Her vision of the white lab coat had been shaped by her postcolonial education, which "paraded the triumphs of dead white men." Later, she looked back and remarked that "the incongruity of a large brown woman in a sea of white men scarcely occurred to me." She imagined that in the United States she would find an intellectual life in science that was free of "national, cultural and social norms of gender, ethnicity, and sexuality." She thought it would be the perfect place for a tomboy from southern Asia. Science, after all, was neutral and objective, so these aspects of her identity wouldn't matter—or so she thought.[2]

Her new home in Durham, North Carolina, seemed friendly. Everyone said hello as they passed on the street. She got used to hearing "How wonderfully you speak English" in a tone of astonishment. Subramaniam would smile and thank people without realizing that their stereotypes about Indian women were following her everywhere. "She came with a stone on her forehead and look what we've done to her," said a professor. "Illegal abortions happen in third world countries; we don't want that here in the industrialized West," said a feminist at a pro-choice rally.[3]

And her academic life was a challenge. In class, where both students and lecturers often viciously shredded one another's work, Subramaniam began to feel increasingly invisible. She had come from a place with few resources and materials, and she was hesitant and careful with expensive equipment, thinking that this was prudent, not unenterprising. Later, she realized that "the world of the alpha male constantly preening, showing off, always looking for opportunities for self-promotion and visibility, was growing uncomfortable and repugnant." She said, "The competitive culture was getting to me." She was no longer in awe of science.

As an antidote, Subramaniam, while continuing her graduate work in evolutionary biology, began taking courses in women's studies. These were a joy, not least because they put her experiences into a historical framework. She discovered that the low number of women in science did not signal a problem with women but with science. She learned that gender norms of the larger society were embedded in the construction of scientific knowledge. She grew to understand that scientific culture was infused with masculinity and see it as a set of practices bound by its roots in western, male, heterosexual culture. And she recognized that graduate school was a way to weed out anyone who did not adhere to these norms and practices.[4]

Evelyn Fox Keller visited Duke while Subramaniam was there, as did Anne Fausto-Sterling. Subramaniam had one-on-one conversations with both women and found them encouraging and supportive. Moreover, they were willing to read her work. She also met Ruth Hubbard. Hubbard's and Fausto-Sterling's writing particularly resonated with her because their work was grounded in biology. In a women's studies course, she read Hubbard's "Have Only Men Evolved?" and started to think differently about Darwin. She read writings by Sharon Traweek and agreed that scientific culture was the "culture of no culture." After two graduate-student friends introduced her to Science for the People, she read every issue she could find.[5]

Once she'd been introduced to these concepts, Subramaniam found it impossible to return to an idealized, apolitical science. Still, for a while her work in evolutionary biology remained separate from her work in women's studies. She knew that there were other feminists who studied science, but she was the only one in her department. Likewise, she knew that there were scientists who considered themselves feminists, but she never encountered one in her women's studies courses. She also felt that most of the feminist writings she was reading focused on human biology. They didn't quite relate to her work in evolutionary biology.

In these years, Subramaniam was working with morning glories; the color variation of their flowers was the topic of her dissertation. In it, she wrote that the study of color polymorphism—the existence of genetic variation and multiple forms—has proven to be an excellent way to explore the forces of natural selection. To gather data, Subramaniam spent a lot of time in a greenhouse observing flowers and plant growth. She was looking specifically at the gene sequence that produced three flower types: dark, light, and white.

She later recalled the exact moment when she first saw the bridge between her two worlds—science and women's studies. It happened when one of her women's studies mentors, Mary Wyer, asked about her doctoral work. Though other colleagues in women's studies had asked the same question, they tended to respond to Subramaniam's explanation with "Oh, how interesting" and move on. Wyer was different. She listened intently and asked more questions. Subramaniam explained her experiments to test the consistency of flower color variation and her effort to explore what evolutionary mechanisms might explain the appearance of variation in flower color. Nodding, Wyer said, "Oh, you work on diversity!" For Subramaniam, this was a "momentous revelation." By suggesting that the concept of variation in her doctoral work in biology was connected to the concepts of diversity and difference in women's studies, Wyer had led Subramaniam to see the connections between these two different disciplines. From then on, her work became interdisciplinary.[6]

Subramaniam undertook a project that would shape the rest of her career: she began to explore how feminist thinking about science could inform experimental, scientific practice. Ironically, the feminist critique of science and a feminist framework had given her the tools to better understand and keep working in science, to find some resolution to the conflict she felt. These new tools would help to not only dismantle the master's house but also build a house that she could live in.

In 1991, during an interview for a documentary film on the women's movement, Rita Arditti was asked to reflect on the past twenty years. She responded:

> I think the consciousness about the importance of science, and feminism and science, has radically changed. I don't think it will ever be the same. When we started talking about those things, we were literally considered crazy. Now the situation is radically different. There is an enormous body of literature written by feminists about the myth of objectivity and what science would be like if a feminist perspective were included.[7]

Back in the 1970s, Arditti had read every single book about women and science that appeared on the shelves at New Words. "Now that's absolutely impossible," she said. "That's been fantastic to watch." These days, she noted, almost every important scientific meeting makes at least a superficial attempt to engage with these issues. She smiled: "We used to go to the AAAS meetings and demonstrate, and do radical disruptions of these meetings. Now, they invite us."

By the early 1990s, Arditti was no longer working in the lab, but she continued to study the role of science in society, especially in relation to women and cancer. It was a good moment to be addressing this issue: cancer was a hot research topic, and many scientists were searching for treatments and cures. In 1991, Arditti co-founded the Women's Community Cancer Project, which focused on the environmental causes of cancer as well as on prevention. Her inspiration was the marine biologist Rachel Carson, who, in *Silent Spring*, had warned of the grave consequences of environmental pollution. Yet Arditti was finding that many scientists preferred to look at science and technology out of context rather than consider their social implications.

Though Arditti acknowledged that women's lives in science had improved, she was scornful of claims that feminism was no longer necessary. She said, "I hate post-feminism. It's a media plot to say that we no longer need to organize. As long as women are alive and struggling with issues in their lives, feminism is very much alive."[8] In her view, there was still much to be done in terms of using feminist thinking to shape scientific research: "I don't think it has permeated the actual research projects. These things take time; these cultural shifts." If feminist activists could reshape society, this, in turn, would reshape science. Much scientific research still did not take women into consideration, and she hoped that feminist science studies would develop new theories, methodologies, and knowledge.

Though many scientists were slow to welcome feminist critiques of science, some of these critiques were nonetheless having an impact on science. A field was beginning to emerge. In 1991, Emily Martin, a young anthropologist at Johns Hopkins University, published "The Egg and the Sperm: How Science has Constructed a Romance Based on Stereotypical Male-Female Roles" in *Signs: A Journal of Women in Culture and Society*. Martin recognized that both popular and scientific accounts of reproductive biology relied on stereotypes of the male as active and the female as passive, and her goal was to shine a bright light on the gender stereotypes "hidden within the scientific language of biology."[9]

Martin's husband, Richard Cone, was a biologist who had become friendly with Ruth Hubbard when he was working at Harvard, and Martin was inspired by her work. She saw Hubbard as a "fierce and fearless" model for young women scholars interested in science. Despite her very famous husband, Hubbard had been forced to climb mountains to achieve legitimacy as a biologist. But "then she built a platform," said Martin, "from which she flew to draw attention to sexism in science, and she welcomed many others to use it for their own flights."[10]

Martin's writings analyzed how scientists and the public use harmful metaphors to describe reproduction—for instance, comparisons to economic production and factories—and emphasized the importance of a feminist analysis of science and how it could have an impact on the course of scientific research. She scoured the human physiology and medical textbooks used in undergraduate science courses at Johns Hopkins and other American universities. The language in these books was telling. For instance, women's biological functions were consistently described in terms of waste and scrap: "debris." "ceasing," "dying," "losing," "denuding," "expelling." The texts also drew on different language to describe female production of an egg and male production of sperm. In one book, Martin read that "whereas the female *sheds* only a single gamete each month, the seminiferous tubules *produce* hundreds of millions of sperm each day."[11]

Sperm was repeatedly presented as production, whereas female eggs were treated as overstocked inventory. One scientist wrote in a newspaper article, "When you look through a laparoscope . . . at an ovary that has been through hundreds of cycles, even in a superbly healthy American female, you see a scarred, battered organ." Martin observed that scientific texts insisted on casting female organs and processes in a negative light. Yet why were scientists describing unfertilized eggs as wasteful? They never used such language for the vast numbers of unused sperm?[12]

In her writings, Martin repeatedly pointed out that scientific language can affect how scientists view what they are observing. In these texts the egg was commonly described as large and "passive." It did not move or "journey" but was "transported," "swept," or "drift[ed]" along. In contrast, the sperm was "streamlined" and active. It "deliver[ed]" its genes to the egg by means of a "strong" tail and great "velocity." Sperm "propel[led] . . . into the deepest recesses of the vagina" with "energy" and "fuel" so that, with a "whiplash like motion and strong lurches," it could "burrow through the egg coat" and "penetrate" it. One text declared that the egg would die unless it were "rescued" by the sperm, a sort of Sleeping Beauty scenario. Although the same text acknowledged that sperm only live for a few hours, it never interpreted this fact as distress.[13] Martin's essay was remarkable because, even as it presented the new research that had changed scientists' understanding of eggs and sperm, it showed that much of the old imagery remained. Scientific descriptions were not shifting to reflect new knowledge. Instead, the old stereotypes were entrenched.

As Martin explained, scientists at Johns Hopkins who were working to develop a spermicide contraceptive had found, to their great surprise, that the "forward thrust of sperm is extremely weak" and that it mostly moves sideways, back and forth. The surface of the egg is what catches the sperm and holds onto it; the two stick together because of adhesive molecules on both surfaces. Nonetheless, in their papers, the researchers continued to treat the egg as the passive partner and the sperm as the active one, if weaker than previously thought. It took the scientists three years after first describing their findings to begin to acknowledge the egg's more active role.[14]

This pattern wasn't limited to the John Hopkins researchers. When scientists in another lab confirmed the findings, they, too, continued to use language that gave more credit to the perilous journey of the sperm. In their experiments, they found that, as the sperm entered the egg, the nucleus of the egg moved swiftly toward the sperm at high speed. But that's not how they described the process in their papers. Another researcher, Paul Wassarman at Mount Sinai Hospital in New York City, also noted a more interactive relationship between the egg and the sperm, identifying a particular molecule on the egg coat that played an important role. Yet his description in *Scientific American* used language that again framed the sperm as the active one, explaining that it "binds tenaciously to receptors on the [egg's] surface."[15]

Martin pointed out that these researchers could have, for example, used the metaphor of two matching lockets to describe the interaction between egg and sperm rather than cling to the old metaphor of a passive lock and an active key. But these new data did not push them to change their language choices or move away from the gender stereotypes that informed their descriptions. She compared this to Darwin's *On the Origin of Species*, which uses concepts from nineteenth-century social ideas about competition to describe the natural world. Once his views of the natural world were accepted as "natural," they were reimported back into the social sphere to justify the existing social structure.

In the case of the egg and the sperm, the sociocultural stereotypes of passive females and active males were imported into the science of fertilization. Once the image was clearly embedded in scientific terms as a description of nature, it was then reimported back into the understanding of what it means to be male and female. Martin concluded that, no matter what personality scientists give to gametes, giving them any personality at all is problematic. In her view, recognizing that we are projecting cultural imagery onto what we study "will improve our ability to investigate and understand nature."[16]

Martin knew that scientists would resist her arguments. So instead of submitting "The Egg and the Sperm" to a scientific journal, she published it in *Signs*, a journal of feminist scholarship. Anne Fausto-Sterling took notice, adding the article to her curriculum and discussing it with her students when teaching developmental biology and fertilization.[17] Ruth Hubbard also read it and afterward asked Martin to give a lecture at the Marine Biological Laboratory at Woods Hole, an elite biological research institution. This was not an easy task. Martin later recalled that most of the scientists in the audience were angry with her, and she was frustrated that they refused to consider her ideas. But she was also encouraged. After the lecture, a young woman sitting at the back of the hall timidly spoke up. With her head bowed and her voice barely audible, she agreed that Martin's descriptions in "The Egg and the Sperm" were accurate and that those assumptions had affected her own scientific work. The woman had been studying lobster sperm in the lab; and when she observed they were not moving, she decided they must be dead.

Later, she learned that healthy live lobster sperm do not move at all. She had been abashed to realize that her expectations about the "vigour" and motility of "male" cells had interfered with her initial scientific observations.[18]

Despite many dismissive responses, Martin did feel that feminist critiques of science were beginning to have an impact. As she traveled the lecture circuit, she learned that more and more students had already read Hubbard's and Haraway's writings. And after one lecture, a few members of the Boston Women's Health Book Collective told her they were planning to edit the next edition of *Our Bodies, Ourselves* to reflect a more accurate scientific account of menstruation and menopause.[19]

In the late 1980s, a group of about ten students working with the biology professor Scott Gilbert founded the Biology and Gender Study Group at Swarthmore College in Pennsylvania. Gilbert had been teaching that a feminist critique of science was one of many important experimental controls in biology. Whenever scientists perform an experiment, they look at what might affect the results. Is the temperature constant? Is the pH steady? What if biologists also routinely asked, Are there any assumptions linked to gender bias? The group believed this view would strengthen biological research going forward.

Group members collaborated on an article, "The Importance of Feminist Critique for Contemporary Cell Biology," published in *Hypatia* in 1988. Like Emily Martin, they argued that "sperm stories" were often treated as mythic quests in which the hero "survives challenges in his journey to a new land, defeats his rivals, marries the princess and starts a new society." They pointed out that, in developmental biology, this active-passive dichotomy had long affected how scientists described sex determination in the embryos of mammals: the embryo actively developing into a male or passively becoming a female. Incorporating feminist critiques could shift such long-held scientific "truths." Because a feminist critique of cellular and molecular biology encourages openness to different interpretations of data, it gives researchers "the ability to ask questions that would not have occurred within the traditional context."[20]

The article included numerous examples of gender assumptions in biological writings. For instance, as cell biology developed as a discipline in the early twentieth century, scientists often described the relationship between the nucleus and the cytoplasm as a marriage. The nucleus was depicted as the masculine ruler of the cell, the brains of the operation. In contrast, the cytoplasm—the thick solution of water, salts, and proteins outside the nucleus but within the cell membrane—was treated as the feminine, fluid, changeable partner.

Not all of these marriage metaphors were identical. In Germany during the 1930s, one dominant theory of the cell was visualized as an autocratic marriage, with the nucleus holding all of the controlling functions and the cytoplasm doing whatever the nucleus demanded. In contrast, the American geneticist T. H. Morgan posited a more democratic model, in which the nucleus and the cytoplasm first consulted together and then the nucleus told the cytoplasm what to do. Notably, as the article pointed out, the husband still made the final decision. (This was also the case in Morgan's own marriage.) By the 1940s, C. H. Waddington, a British geneticist, was suggesting that the nucleus did not dominate the cytoplasm but worked in partnership with it. (He was married to a successful architect and respected women as intellectual equals.)[21]

The article also made reference to Ernest Everett Just, the subject of Kenneth Manning's *The Black Apollo of Science*. In the 1930s, when Just was writing about cell biology, most of his colleagues believed that the nucleus drove the life of the cell. Just, however, wanted to shift attention to the reciprocal interaction between the nucleus and the cytoplasm. He believed that the ectoplasm, the outer part of the cytoplasm that formed the surface of the cell, played an important role in communicating with the external environment of the cell. His views were similar to those of E. B. Wilson, often called the first American cell biologist. Study group members argued that this interpretation of the cell reflected how Just viewed not only the cell but also his own less controllable experiences with male-female relationships. [22]

The article concluded that all of these views of nuclear-cytoplasmic interactions reflected personal views of male-female interactions. Nonetheless, in the late twentieth century, most cell biologists were still using language that equated the nucleus with a dominating male and cytoplasm with a passive female, and this perception was affecting the direction of research. Because many see science as objective, this view of the cell was also supporting the social views of sex, gender, and marriage that had been imposed metaphorically on the cell.[23]

The Biology and Gender Study Group took their analysis beyond biology to illustrate how the metaphor of the interaction between the active sperm and the passive egg had also influenced organic chemistry concepts. The metaphor of penetration and entry often appears in organic chemistry lectures and texts. For instance, collisions between two molecules that lead to the formation of new compounds are depicted sexually, with a small active molecule "attacking" a larger, passive, heavier one. The article quoted a text that described the alkene bond as "being 'ripe for plucking' by an approaching electrophile." But why, asked the writers, is it necessary to gender molecules and chemicals?[24]

In 1985, Scott Gilbert, one of the members of the study group, had published *Developmental Biology*, an influential textbook that discussed how the feminist critique of science was strengthening the discipline.[25] Gilbert had both a PhD in biology and a master's degree in the history of science, the latter under the supervision of Donna Haraway, which had no doubt influenced his perspective. Emily Martin later recalled that, by the early 1990s, she was interacting with young med students who had been assigned *Developmental Biology* in college. Given their growing awareness of feminist critiques of science, she could tell that it was having an impact.[26]

CHAPTER 16

BUILDING TWO-WAY
STREETS

In August 1991, a journalist phoned Anne Fausto-Sterling to ask if she'd heard about a new study by Simon LeVay claiming that homosexual and heterosexual men's brains were different. Could she read the study and comment on the new findings? Fausto-Sterling read the article and spent a long time talking to the reporter about it. Then she got another phone call. And another. The time she spent fielding press calls about this article grew from hours into days.[1]

Fausto-Sterling was frustrated. The public discussion was not only circling around homosexuality but also reigniting old assumptions about gender and sex differences. The logic was that men and women have different brains, so gay men might have brains that look more like women's and lesbians might have brains that look more like men's. Apparently, because Fausto-Sterling had written about sex differences and cognition in *Myths of Gender*, she had become the go-to scientist for comment. Yet she knew this was not a scientific debate but a discussion about social inequality. In her opinion, the question in the media really was "Should gay people have full civil rights? Should they be treated as if they have a disease?" It fell into the same category as "Should little girls be expected to perform well in mathematics and become engineers?" Answering the same questions over and over again made her weary.[2]

The United States had spent more than ten years under the conservative leadership of Ronald Reagan and George H. W. Bush, and these phone calls were part of the current backlash against the women's movement and gay activism. So Fausto-Sterling suggested to her publisher that they put out a second edition of *Myths of Gender*. She added a new chapter examining recent research on sex differences, homosexuality, and the anatomy of the human brain and an afterword that offered updated analysis.[3]

Fausto-Sterling spent the decade after first publishing *Myths of Gender* immersed in the field of science and technology studies. But in the early 1990s, she was asked to teach a course at Brown about developmental genetics in comparative vertebrate embryology. She had first taught the course in the early 1970s and had periodically offered it again over the next fifteen years, but there had been a long gap since she'd last led it, and a lot had changed. Originally, she'd focused on how groups of cells take on three-dimensional shapes that grow to form an embryo and then a baby. Students would use information from lectures and readings to inform their work in the lab, where they were examining embryos of frogs and chickens at different stages of growth after fertilization. But now she had a very different view of science; and when she moved back to teaching the embryology course, she came into conflict with some of her long-time colleagues in biology.[4]

Fausto-Sterling realized she was having difficulty communicating with her colleagues and students. For the first time in her life, she was getting terrible student evaluations. They told her she didn't seem prepared or seem to care about the class. Fausto-Sterling found these comments painful because she *did* care. Why were people unable to understand her point of view? She began to realize that she could no longer teach science in the same way, that she would have to change the course dramatically.[5]

In the 1970s, during a lecture on the development of the central nervous system in an embryo, Fausto-Sterling would include a description of the events that led to the formation of the neural tube. Now, after years immersed in feminist theory and science studies, she saw the neural tube differently: it was embedded in an interconnected set of environmental, medical, historical, and social conditions. She couldn't mention embryonic neural tubes without mentioning birth defects such as spina bifida, which would then lead her to a discussion of health care.

Again and again, her colleagues at Brown said, "It's fine if you teach about the social impact of embryology as long as you still teach the anatomy." Each time they made that point, she became more frustrated. Did they really imagine she would teach a course on developmental anatomy without teaching anatomy? Or were they trying to downplay the importance of related critiques?

Fausto-Sterling turned to the work of Bruno Latour, a French philosopher known for his work in science and technology studies, to make sense of her situation. In his writings, he pointed out that, during the Enlightenment, a division had occurred between those who studied human culture and those who studied nature. This separation, he argued, was false. After spending time with Latour's explanations, Fausto-Sterling realized that her colleagues' belief in this separation was at the heart of the communication problem.[6]

Searching for a solution, Fausto-Sterling reviewed Emily Martin's study of scientific knowledge about the immune system. Martin had found that most biology researchers see the human body as self-contained and separate from the outside environment, whereas many nonscientists see it as intimately affected by the environment. The second view is the minority opinion among scientists. Yet at some point, as Martin wrote, it might become the majority one, the view that disseminates "the facts" about the body. This is a common trajectory; but when it happens, scientists often don't remember that they once rejected what they now embrace. Fausto-Sterling had seen this pattern in other contexts. First, the work and writings of feminists and people of color had been rejected. Later, those same ideas were embraced and co-opted.[7]

Eventually, Fausto-Sterling found a new model for teaching science, one that she called "science in social context," and she asked her department chair if she could develop a new course called "Embryology in Social Context." The mechanics and genetics of the neural tube in an embryo were still at the center of the class, but she added concerns about birth defects and the occupational exposure, nutritional problems, and contaminated drinking water that contributed to changes in development. She related these to issues of race, class, and gender. The course employed active student learning instead of a lecture-only format. Students engaged with imperfect scientific knowledge about neural tube defects; and different ethical questions arose regarding amniocentesis, genetic counseling, and abortion. By redesigning her course, Fausto-Sterling "let the cultural complexities of scientific knowledge become visible in the classroom." She realized that, "by permitting our students to grapple with these [issues]," it was possible to reinvigorate science education. In the process, many students—"women, minorities, socially concerned

students from a variety of backgrounds"—suddenly viewed the science classroom as a compelling place to be.[8]

Now Fausto-Sterling began to think about writing a new book. What if
biology itself were socially constructed? How could she rethink concepts
of sex, gender, hormones, and genitalia? Throughout the 1990s, she delved
into science studies and scientific research, and in 2000 that work came to
fruition in *Sexing the Body*.

By the early 1990s, Fausto-Sterling and her husband were divorced and
she was dating Paula Vogel, a playwright and a professor of creative writing at
Brown.[9] In the introduction to *Sexing the Body*, Fausto-Sterling wrote that, as
someone who had lived part of her life as an "unabashed heterosexual, part as
an unabashed lesbian, and part in transition," she was open to theories of sexuality that allowed for the development of new patterns of behavior over time.[10]

She recognized that society and the medical community needed to understand that the relationship between masculine and feminine is not binary and
that gender differences fall on a continuum, both culturally and biologically.
So, in 1993, she published the article "The Five Sexes: Why Male and Female
Are Not Enough" in *The Sciences*. She knew the piece would be provocative
but was surprised at the extreme reactions from the Catholic church and
from other academics. In a much-later follow-up article, "The Five Sexes
Revisited" (2000), she said she had written the first one with "tongue firmly
in cheek" and that she had not intended to promote a specific number of
sexes. Nonetheless, she had no regrets about her overall point. It had given
her pleasure to think that she had played a small role in encouraging society
to think beyond the framework of two diametrically opposed sexes.[11]

In the 1990s, Banu Subramaniam was juggling her dissertation research in
biology with her courses in women's studies. She wrote about this both-and
balancing act in "A Contradiction in Terms," published in 1998 in the *Wom-
en's Review of Books*:

> Almost a scientist, yet a feminist; almost a feminist, yet a scientist, almost a
> resident, yet an alien; almost an alien, yet a resident; almost a heterosexual,
> yet homosexual; almost a homosexual, yet heterosexual; almost an Indian,
> yet American, almost American, yet an Indian; almost an outsider, yet
> inside; almost an insider, yet outside. . . . Almost there, but never quite. A
> life held captive in oppositions. How did I find myself in this tantalizing,
> much celebrated place, the home of the oxymoronic feminist scientist, this
> magical yet insane place, . . . nowhere, yet everywhere all at once?[12]

In 1992, she'd read an article by Anne Fausto-Sterling, "Building Two-Way Streets: The Case of Feminism and Science," published in the *National Women's Studies Association Journal*, which had helped her begin to understand herself as an "oxymoronic feminist scientist." The piece opened with a quote from Evelyn Fox Keller describing the challenges that face women scientists. According to Keller, if society viewed a woman scientist as existing between a rock and hard place, then being a feminist scientist "ma[de] matters worse." She bemoaned the lack of an intellectual community: there was no fully welcoming place for her among either feminist scholars or scientists.[13]

In the article, Fausto-Sterling described the experience of delivering a lecture at a women's studies event at a university. At the podium, a colleague introduced her: "I'm not going to talk about Professor Fausto-Sterling's scientific work because I can't pronounce any of the words." What upset Fausto-Sterling about the situation was not that the women present were unfamiliar with her research but that so many had been socialized to think that this disconnect from science was okay. Women scholars had no problem reading aloud the name of an Italian city-state or a reference in Latin, even if the words were unfamiliar. Yet she regularly encountered highly educated women who wouldn't even attempt to read the titles of her scientific articles. Women did not have to buy into this attitude, and it set a terrible example for students.

Fausto-Sterling had had equally frustrating conversations with her science colleagues. In the hallway or at the lunch table, she would talk about scientific events but would feel the need to filter out references to science as a social activity or as situated within a historical context. Most scientists rejected a contextual approach, and that rejection was socially accepted.[14]

These two problematic scenarios—one of women not engaging with science, the other of scientists not engaging with their own context—were being reproduced again and again throughout the educational system and throughout society. Women don't need to understand science. In fact, science is an inappropriate place for them. And scientists don't need to understand their social context. In fact, that context has no place in science. "It is, I am afraid, we—those engaged in gender studies and in the sciences—who bear a significant portion of the blame for these inaccurate points of view," wrote Fausto-Sterling. She noted that most women's studies majors do not have a science requirement and that too few science faculty members teach science in its social context.

Fausto-Sterling's essay was an eye-opener for Subramaniam. She knew that most feminists in women's studies saw scientists as the problem. "They have to learn how gender is constructed. They have to understand the history of misogyny and racism in the sciences." But Fausto-Sterling's article helped Subramaniam see that feminists were partially to blame as well. If science had developed as a world without women, then women's studies had developed as a world without the sciences.[15]

Subramaniam believed that the feminist critique of biological determinism was vital to feminist thinking. However, she lamented that that critique was the only space for science in the women's studies classroom. For many feminists, she realized, eugenics, the Tuskegee Study, and Nazi medical experiments were their only visions of science. Subramaniam didn't just want to use her feminism to critique science; she also wanted science to inform her feminism. She wanted a women's studies program that welcomed the sciences and scientists.[16]

Evelynn Hammonds had foreshadowed Fausto-Sterling's concern about this issue in an interview in 1986. She spoke of being relieved after some of her friends read Donna Haraway's article "Cyborg Manifesto" and wanted to talk about it: "The fact that they wanted to read a paper that had a lot of science in it was like 'oh thank you.'"[17] But Fausto-Sterling's article was more than a theoretical discussion. It also offered a concrete example of why this two-way exchange was important. The Human Genome Project, wrote Fausto-Sterling, had major implications for society and was being promoted as a significant contribution to human welfare. Yet many feminists weren't paying attention.

In the late 1980s, scientists had started raising funds for what they described as a plan to map all human genes. The initiative, known as the Human Genome Project, quickly grew into a $3 billion, fifteen-year-long enterprise. Several feminist scientists were engaged with the project, including Ruth Hubbard and Evelyn Fox Keller. Now Fausto-Sterling was suggesting that it would be beneficial for a wider group of feminists to engage with this and other scientific endeavors.[18]

As the Human Genome Project was getting underway, Ruth Hubbard, Evelyn Fox Keller, Anne Fausto-Sterling, and other feminist scientists were reconsidering the concept of the gene and of DNA as a master molecule. They took exception to the traditionally male-centric assumptions that contributed to such a concept.

In the first half of the twentieth century, genes had been described as abstract hypothetical units. It was only after the explanation of the double helix structure of DNA, and a better understanding of the mechanism of replication in the 1950s, that genes began to be characterized as concrete entities. In the following decades, as molecular biologists learned more about how sequences of DNA (that is, genes) influence the essential processes of living organisms, two of these researchers, James Watson and Francis Crick, began to refer to DNA as the *master molecule* of life, suggesting that it was the center of control in the cell and the organism. Many textbooks still use this term. The pair also introduced the hierarchical notion of *central dogma*, stating that DNA makes RNA, which, in turn, makes a protein in a linear process—again, showing that DNA is in control.

As early as the 1970s, Hubbard had critiqued the notion of central dogma, arguing that there was a more interactive relationship among DNA, RNA and proteins. And she had also lambasted Watson for saying, "The best home for a feminist is in another person's lab."[19] In the 1990s, Hubbard's *Exploding the Gene Myth* (1993), written with her son Elijah Wald, as well as Keller's *Refiguring Life: Metaphors of Twentieth Century Biology* (1995) and *The Century of the Gene* (2000), contributed to critiques of the central dogma as inadequate. Bonnie Spanier, who had studied with Hubbard in the late 1970s, published a book about gender and ideology in molecular biology, *Im/partial Science* (1995).[20] All of these books contributed to new thinking about genes and genetics that acknowledged the two-way flow between biology and the environment.[21] With the Human Genome Project having a large impact on the public understanding of science, Anne Fausto-Sterling was suggesting, in her essay, that a broader group of women needed to give attention to these issues.

The editors of the *National Women's Studies Association Journal* invited four women scholars to comment on Fausto-Sterling's article "Building Two-Way Streets": Hubbard, Sandra Harding, Nancy Tuana from the University of Texas at Dallas, and Sue Rosser from the University of South Carolina. In her response, Hubbard agreed that both science and women's studies would benefit if a greater number of feminists had scientific training. However, she also pointed out that many college-level science classes are boring because they focus on the repetition of facts. These courses would attract a wider variety of students if they focused on asking exciting questions about the world.

Hubbard also agreed that scientists do not generally welcome colleagues who ask questions about the social context of science or critique the way in which it is practiced. She understood that research scientists often had to "keep their feminist and sociological interests in the closet" and knew that young scientists working toward tenure were in a difficult position if they were interested in feminist science studies.[22]

Harding saw Fausto-Sterling's article as a call to all feminists who were not scientists, including herself. But she also drew a link between women's disinterest in science and science's disinterest in its own social histories. She proposed that, in the same way that English courses include both writers and literary critics, science courses should include both experimental scientists and social analysts of science and its history. In her view, as soon as social movements believed that science was working with them rather than against them, their interest in science would increase. This was already happening, she said, in health sciences and ecology.[23]

Fausto-Sterling was happy that her colleagues had taken the time to discuss her article. In her own response, she asked a new question: how can we best encourage students to construct a new form of scientific knowledge, one that would have a strong methodological base and would serve the basic needs of the majority of the people? Although Science for the People was no longer in operation, this was the kind of question that the organization had long been asking. Fausto-Sterling urged everyone to keep working diligently toward this goal. "It isn't a very exciting way of thinking about social change," she acknowledged. "No barricades to storm, just plodding and arguing, one tiny step at a time."[24]

CHAPTER 17

WHERE ARE THE WOMEN OF COLOR?

In the early 1990s, Evelynn Hammonds often wore a starched white shirt, tight jeans, and red cowboy boots. On the day she was getting ready for her PhD exam, a fellow student told her that she would likely fail. When she asked why, he pointed to her red boots and her dreadlocks. "If they fail me for that," responded Hammonds, "I don't actually care."[1]

Not only did Hammonds pass her exam and finish her doctorate, but she also did well in the job market, interviewing at Bryn Mawr College, Ohio State University, and the University of Chicago. When she called her friend Robin Kilson, an African American assistant professor of history at MIT, to ask about teaching at Bryn Mawr, Kilson suggested that she also apply at MIT. Hammonds was hesitant, but then Ken Manning also encouraged her.

Hammonds had never imagined she would go back to MIT. She was pretty sure she would end up at the University of Chicago. But her interview at MIT went well, and the next thing she knew she was being offered a position as assistant professor of science, technology, and society. When she moved into her new office, Hammonds felt that her life had come full circle. Back at MIT, she was taking care of unfinished business: she immediately started teaching a course on race, gender, and science.[2]

For years, Hammonds had been delivering papers at the annual conference of the History of Science Society (HSS), but her talk in November 1993 was her first as Dr. Evelynn Hammonds. Her topic was "Race, Gender, and the History of Women in Science," and it included a discussion of the woman who had long captured her attention—Roger Arliner Young.[3] This was the first time that anyone had ever given a presentation on Black women scientists at the HSS. As members of a traditional, largely white organization, HSS feminists generally focused on the history of white women scientists. While they supported Hammonds, she felt that they couldn't relate to the issues that mattered to her, and she was angry that her work was seen as marginal. In her presentation, she tried to show her colleagues, friends, and mentors that, as historians of science, they were actively erasing the experience of African American women and contributing to their continued marginalization. Although Hammonds did not want to make the talk personal, she did want to make the point that the obstacles Young faced were also relevant to herself.[4]

In these years, Black women historians such as Darlene Clark Hine and Elsa Barkley Brown were raising a challenge to white women historians, and Hammonds wanted to do the same in the history of science. During her research, she had discovered a novel that was loosely based on Young's life.[5] In the novel, the Young character, a woman scientist studying for her PhD, lives with several white women who are fellow students, who have hired her to do housework as well, so her position in the household is radically different. When she comes home from the lab, she changes her clothes, cleans the house, and serves them. In her presentation, Hammonds brought up the novel as a way to share these realities with her colleagues. Afterward, when they came up to speak to her, they were all very nice about her talk, but Hammonds still didn't feel that they were ready to enter deeply into her concerns.[6]

In the mid-1990s, there were so few African American scholars in the history of science that many of them could fit into the same car. One of these scholars was Vanessa Northington Gamble, who'd been a good friend of Hammonds since the Hampshire College conference years earlier. Gamble had both a medical degree and a PhD in the history of science, and she was hosting a conference at the University of Wisconsin on the history of race and medicine. Hammonds attended; and after the first day had ended, she, Gamble, Ken Manning, and Gerard Ferguson (another African American scholar) all piled into Gamble's car to head to dinner. Gamble was a fast driver, and Manning, alarmed, said, "You're going to kill us!" She glanced over at him, smiled, and commented that an accident would wipe out half the Black people in the history of science.[7] Despite their small numbers, however, the impact of this small cohort of scholars was foundational in the 1990s. They initiated important work in race and slavery studies, race and the history of medicine, as well as the history of racism in science.[8]

In October 1991, Anita Hill testified before the Senate Judiciary Committee about her allegations of sexual harassment against Clarence Thomas, who had been nominated to become a Supreme Court justice. Hammonds watched the hearings, which seemed to her to be "the most shameful display of misogyny" she had ever witnessed.[9] She couldn't believe what she was watching on national television. Hammonds was so incensed by how Hill was being treated that she decided to write an essay. As she worked, she looked up at a poster on her bedroom wall, which advertised a radio show with Audre Lorde. To Hammonds, the words printed on the poster, "Your silence will not protect you," stood out vividly.[10]

In her essay, "Who Speaks for Black Women?," published later that year in *Sojourner*, Hammonds argued that the Senate proceedings had revealed the failure of a Black feminist movement in the country. Though many pundits were suggesting that Hill had a "duty to her race," Thomas apparently only had a duty to himself. Hammonds knew that Black feminists had produced an impressive amount of theory and analysis about how race and gender affected Black women's lives. So why weren't they presenting this at the hearings and to the media? She called for the founding of a national Black feminist organization, and she was one of more than 1,600 Black women who published a letter of support for Hill in *The New York Times*.[11]

In a letter, Audre Lorde, now very ill, congratulated Hammonds on the essay, and Hammonds replied, saying that Lorde's own words had given her the courage and offering prayers for her friend's health. But less than a year later, on November 17, 1992, Lorde died. Her passing was a great loss to Hammonds, yet the power of their friendship lived on.[12] Spurred by the Hill hearings and Lorde's death, she continued thinking about how to bring Black women feminists together. With Robin Kilson, she started planning a conference, "Black Women in the Academy: Defending Our Name, 1894–1994." They scheduled the gathering for January 1994 and hoped to attract several hundred Black women scholars. In the end, they drew more than 2,500 people and featured three prominent plenary speakers: Angela Davis, Lani Guinier, and Johnnetta Cole. It was the first time that MIT had hosted such a conference, making it an important landmark event.[13]

The Institute for Advanced Studies at Princeton University had offered

Hammonds a postdoctoral fellowship, and in 1994 she spent time there with white feminist scholars and was also able to visit many Black women scholars and friends in New York City. During this time, she wrote her most famous essay, "Black (W)holes and the Geometry of Black Female Sexuality," which has since been reprinted in many anthologies and is often required reading in women's studies programs.

Hammonds opened the essay with the words of Lorraine O'Grady:

> The female body in the West is not a unitary sign. Rather like a coin, it has an obverse and a reverse: on the one side, it is white; on the other, not-white, or prototypically black. The two bodies cannot be separated, nor can one body be understood in isolation from the other in the West's metaphoric construction of "woman."[14]

Hammonds went on to address racialized sexuality, Black women's bodies, and Black lesbian sexuality in ways that she had never done before. She took readers back to eighteenth-century European colonial elites and scientists and carried them forward to Kimberlé Crenshaw's concept of intersectionality. At the time, queer theory was a new influence, and Hammonds drew on it to chart the history of negative stereotypes about Black women and their sexuality. She borrowed the concept of a black hole from physics, noting that it is often used to describe a void but that it is, in fact, not empty but dense and full.[15]

Hammonds had already been thinking about the need to convene a forum in which she could read and discuss race and science. She also knew there was a need for more complex stories about what was happening to women of color in science. Everett Mendelsohn, who had been one of her PhD supervisors at Harvard, encouraged the idea. So when she returned to MIT, she created a monthly gathering called the Workshop on the History of Race in Science, Medicine, and Technology, which would look at intersectionality and how gender intersected with race in science. Hammonds spread the word and brought pizza.[16] And people came.

The MIT Workshop on Race and Science ran for a decade. Graduate students who participated went on to become full professors who wrote books about race and science from a feminist perspective. The scholar Jenny Reardon attended as she was working on her book, *Race to the Finish*, about the Diversity Project in the age of genomics. In her preface, she wrote that, without the support of Evelynn Hammonds, she would not have embarked on the book. Another regular was Lundy Braun, the author of *Breathing Race into the Machine*, about how racism had been embedded in one medical instrument in the antebellum South. In *Body and Soul*, about the Black Panther Party and its campaign against medical discrimination, the scholar Alondra Nelson acknowledged Hammonds as an important mentor and expressed her gratitude for the workshop.[17]

Anne Fausto-Sterling also attended. She and Hammonds spent countless hours talking about issues of gender, race, and science, and the gatherings pushed her to think about writing about Sarah Baartman, a South African woman who had lived in the early 1800s. In the end, Fausto-Sterling decided not to write the book but to include her thoughts in a long essay titled "Gender, Race, and Nation."[18]

At Duke, Banu Subramaniam heard about the workshop and wished she could be part of it. But she was finishing up her PhD so couldn't travel to Cambridge for the meetings. Then, in early 1994, Mary Wyer, a professor of women's and gender studies, invited Hammonds to Duke to speak to women of color in science and engineering. Subramaniam was excited to meet her, and their relationship would grow over the years. Eventually, they would publish an article together.[19]

After earning her PhD in evolutionary genetics, Banu Subramaniam took a position as the director of the women in science program at the University of North Carolina at Chapel Hill. A few years later, she accepted a job at the University of Massachusetts in Amherst, which gave her the chance to travel to Cambridge to see Hammonds and take part in Boston area events.

Meanwhile, other feminist scholars were starting to recognize that an understanding of race and racism was important to feminist science studies. In *Primate Visions*, Donna Haraway examined how race and gender had shaped the field of primatology. In her essay "Race and Gender: The Role of Analogy in Science," Nancy Stepan explored the intersection between gender and race in nineteenth-century science. In her edited volume *The "Racial" Economy of Science*, Sandra Harding attempted to build a bridge between growing feminist scholarship around science and her own recognition that science was Eurocentric.[20]

In the late the 1990s, the sociologist Patricia Hill Collins published "Moving Beyond Gender." In her opinion, the feminist analysis of science had moved in important directions during the past fifteen years, yet she agreed that most feminist theory continued to reflect the viewpoints of white middle-class women from North America and western Europe. Collins reemphasized that gender is constructed by race, class, ethnicity, and other factors and explored how the feminist analysis of gender, race, and scientific knowledge could "benefit from closer association with emerging scholarship on intersectionality."[21]

Collins applauded the writings of Anne Fausto-Sterling and the contributions to Harding's The "Racial" Economy of Science. But she was less complimentary about three other recent anthologies about the feminist analysis of science. One of these was Feminism and Science, edited by Evelyn Fox Keller and Helen Longino. Even the authors had noted in their introduction that all of their contributors were "white and Western."[22] Collins saw this gap as a major flaw.

In 1995, Banu Subramaniam and Evelynn Hammonds both attended a conference on "The Women, Gender, and Science Question" at the University of Minnesota. The gathering was an attempt to bring together scholars who worked on women's participation in the sciences and those from the field of gender and science as there was still no clear integration between the two disciplines. Unfortunately, according to Hammonds, the conference offered sessions about women in science that addressed equity issues and how to encourage more girls to study science. And there were sessions about gender and science, including the philosophy of science, sociology, and the history of science. But there was very little discussion across these topics.[23]

Subramaniam was frustrated. "As someone who wanted to continue doing experimental biology, I was starving for locations where everyone was at the same table so we could talk about the relationship between these subfields." But that didn't happen at this conference. "These cross-linkages never got made, officially or unofficially, because ultimately everyone went to their subspecialty and the different groups were never in the same room."[24]

The women agreed that some of the individual sessions were good, among them Keller's plenary talk about gender and science. But their overriding impression was that the two different streams—women in science and feminist science studies—did not overlap. Five years had passed since Hammonds's essay about conflicts and tensions between women scientists and feminist science studies, and little had changed.[25]

Robert Merton's 1968 article "The Matthew Effect in Science" exposed a common pattern in which famous, largely male scientists receive more awards, funding, and recognition than lesser known but equally accomplished scientists. Merton was using the term Matthew effect to describe a kind of halo around well-known male scientists who are celebrated for findings and achievements that they have not achieved on their own or at all.[26]

Twenty-five years later, in her 1993 article "The Matilda Effect in Science," Margaret Rossiter coined a companion term, the *Matilda effect*, to describe an equally common pattern, in which the work of women scientists is not acknowledged but attributed instead to their male colleagues. She named the term after Matilda Joslyn Gage (1826–98), a feminist and suffragist from upstate New York. One of Rossiter's examples was Rosalind Franklin. Another was the German scientist Lise Meitner, who worked for years with Otto Hahn and in 1938 identified the nuclear splitting process, calling it nuclear fission. Yet Hahn alone won the Nobel Prize in 1944.[27]

"If unmarried female collaborators often receive less credit, the pattern is even more pervasive among collaborative married couples," wrote Rossiter. She specifically mentioned Ruth Hubbard and George Wald and the work they had done, both independently and collaboratively, on the biochemistry of vision between the 1940s and the 1960s. After Wald won the Nobel Prize in 1967, all of Hubbard's previous work was retrospectively attributed to him. "For some possibly threatened scientists, marrying one's collaborator may be a strategy for undercutting a serious rival in the race for recognition," wrote Rossiter.[28] Hubbard would probably not have agreed with Rossiter on Wald's motivation for marrying her, but she did once tell Patricia Farnes, a professor of medical science at Brown, "I didn't win the Nobel Prize for George. A lot of us, including George, worked together. Laboratories should get prizes— not people."[29]

CHAPTER 18

SCIENCE STUDIES, SCIENCE WARS, AND THE SOKAL HOAX

In the early 1990s, a number of scientists were expressing concerns about the social study of science, also known as science and technology studies, especially in relation to the critique of modern science. As Sandra Harding said, "no one likes to be studied." A conservative breeze was blowing in politics, and it was having an impact on academia, where scientists were criticizing what they perceived as leftwing points of view, including feminism. In 1994, the biologist Paul Gross and the mathematician Norman Levitt published *Higher Superstition: The Academic Left and Its Quarrels with Science*, which countered the arguments that science was not objective and that it was influenced by society and politics. The book's longest chapter, "Auspicating Gender," focused on the feminist critique of science and included the cringe-worthy statement "Sexist discrimination, while certainly not vanished into history, is largely vestigial in the universities. . . . The only widespread, obvious discrimination today is against white males."[1]

In 1996, in reaction, the journal *Social Text* released a special issue to engage directly with these questions. Ruth Hubbard and Sandra Harding were among the contributors, as was a physicist at New York University named Alan Sokal. Sokal chose to submit a parody article titled "Transgressing the Boundaries: The Transformative Hermeneutics of Quantum Gravity." His intent was to prove that the editors of a science studies journal would accept anything if it were packaged as a critique of science. And, in fact, these editors were happy to include a critique of theoretical physics.

After Sokal revealed his hoax, a number of scientists came to his defense, among them E. O. Wilson and Richard Dawkins. Likewise, several philosophers of science and historians of science condemned the hoax, suggesting that he was trying to paint all critiques of science with the same brush and reject the entire field of science studies. The ensuing years of debate came to be called the Science Wars. In their own response, Gross and Levitt declared that the Sokal hoax "brought into the open a widespread reaction, years in the making, against the sesquipedalian posturing of postmodern theory and the futility of the identity politics that so often travels with it."[2]

Without Science for the People, there was no organizational response against the hoax. This allowed the media to assume that all scientists were opposed to science studies, which was not the case. Instead, scholars such as Hubbard, Keller, and Harding had to defend themselves on their own.

Hubbard's own article, "Gender and Genitals: Constructs of Sex and Gender," appeared in the *Social Text* special issue.[3] But once the fighting started, she stepped back from the fray. She was in her seventies and was starting to feel her age. Moreover, her husband George Wald, now in his nineties, was not well. He had entered what Ruth called "the Tunnel" and she wasn't sure how long they had left together. (Wald died in 1997).[4]

At this moment in her life, Hubbard could not become involved in academic jousting. Yet her existing paper trail spoke for her. Nearly ten years earlier, in an article for *Science for the People*, she had written, "The pretence that science is objective, apolitical and value-neutral is profoundly political because it obscures the role that science and technology play in underwriting the existing distribution of power in society."[5] She had long been a staunch defender of the feminist critique of science; and as the Science Wars heated up, her supporters came to her defense. In 1994, in a long article in the *New York Review of Books*, Richard Lewontin celebrated Hubbard and her scholarship. "No one," he said, "has been a more influential critic of the biological theory of women's inequality."[6]

Harding and her standpoint theory were also under attack, but, like Hubbard, she stayed out of the fray. She decided that engaging in the fight would only give it more oxygen. Instead, she continued to write about her theories on their own merit. After each attack, she would not engage directly but move on, though sometimes she would articulate her counter-position in a subsequent essay.[7]

Harding later said that Gross and Levitt and those who agreed with them attacked feminism and the feminist critiques of science because they knew they would likely receive support from women scientist colleagues who were uncomfortable with the critiques. Harding also felt that feminists were an easy target. "At the time," she said, referring to herself and Keller, "we weren't even tenured." As it happened, the Science Wars were relatively brief—in Harding's view, because postcolonial critiques soon began making some of the same arguments against modern western science. It was easier to attack young feminists than senior postcolonial scholars.[8] Nonetheless, the Science Wars and Sokal's prank did long-term damage by giving rise to anti-intellectual reactions in the media and among the public. As the commentator Katha Pollitt, pointed out, "now people who have been doing brilliant, useful work for years in the social construction of science . . . will have to suffer, for a while, the slings and arrows of journalists."[9]

The Science Wars arrived at a bad time for Evelyn Fox Keller. In 1992, she had been thrilled to accept a permanent position in the science, technology, and society program at MIT. After years spent shifting from university to university, she was happy to finally receive a professorship and was pleased that Evelynn Hammonds would be her colleague in the department. Moreover, soon after accepting the job, she learned that she had been named a MacArthur Fellow. The award brought her prestige but also money. Yet it also brought with it tensions among her MIT colleagues.

The Science Wars had ushered in an atmosphere of defensiveness among the scientists at MIT, and it was in this environment that Keller gave her first formal talk as a newly appointed professor and MacArthur fellow. The room was packed for her lecture, and the discussion that followed degenerated into a personal attack on Keller as well as on the more general notion that people who were not practicing science might dare to critique it. Keller's advocates were also vociferous. At one point, Thomas Kuhn, a historian of science and the author of *The Structure of Scientific Revolutions*, yelled at an outspoken physicist, "What do you know about objectivity?"[10]

As the Science Wars showed, feminist scientists and philosophers of science were under threat in the 1990s. Yet at the same time, they were also broadly ignored by many in the academy—not only scientists but other science studies scholars. Donna Haraway was exasperated by this pattern. So in 1994, at the annual conference of the Society for Social Studies of Science, which met

jointly in New Orleans with the History of Science Society and the Philosophy of Science Association, she delivered a scathing critique of Bruno Latour,
the French philosopher well known for his work in science studies. Haraway
was already acclaimed among feminists for her articles "A Cyborg Manifesto"
(1985) and "Situated Knowledges" (1988) as well as her book *Primate Visions*
(1989).[11] But her talk in New Orleans, "Never Modern, Never Been, Never
Ever: Some Thoughts about Never-Never Land in Science Studies," had an
impact on a much broader group, many of whom were not feminists.

Latour was a speaker at the same gathering and was listening as Haraway
spoke. He had recently published *We Have Never Been Modern*, in which he
argued that the rise of science had led to a false dichotomy between nature and
society: modern thinking had come to see earlier thinking as primitive and pre-
modern. While Haraway agreed with many of Latour's points, she called attention to the fact that a great deal of feminist literature had already made them.
She mentioned Carolyn Merchant's *The Death of Nature* (1980), for instance,
and called attention to the writings of Rita Arditti and Evelyn Fox Keller. All
would have supported Latour's thesis had he chosen to cite them.[12]

Haraway and Latour were friends and colleagues, but they had often tus-
sled in discussions. Now, in this public forum, she declared, "For a long time
now you have been completely tone-deaf to intersectional feminist theory, to
feminist science studies, to non-Western thinking, in spite of an occasional
example in your writing." She accused him of drawing only from masculinist,
individualist, scholarly sources and pointed out that he often used war as a
fundamental trope in his writings about network-based knowledge making
and science. "You can't do this work," said Haraway, "unless you engage with
the other social movements that are at the root of science studies, most certainly including intersectional feminist science studies—the kind that Patricia Hill Collins, and myself are doing. And Sandra Harding and Evelynn
Hammonds, and, and, and. . . ." Latour did not cite any of these sources, she
said, and people were getting angry about it.[13]

Although Haraway's tone was humorous, her point was serious. Both
Anne Fausto-Sterling and Evelynn Hammonds were in the audience that
day, and, as Fausto-Sterling recalled, "Donna took Bruno apart. He was
somewhat chastened after that." The listeners were stunned, but the effect
of Haraway's talk went far beyond the immediate audience. Her speech was
a turning point, marking the moment when feminist work in science studies
began to receive greater recognition.[14]

Nonetheless, the Science Wars continued. Keller was bruised by several articles in the *Boston Globe* that pointed to her, Harding, and Latour as leaders in feminist studies and science studies. The academic backlash was harsh. Keller recalls that she was told that it would be best for her credibility to disassociate herself from Harding's work, though, in fact, she had had numerous disagreements with Harding over the years. She felt trapped: she didn't want to cut ties with Harding, nor did she not want to be labeled as "anti-science."[15] But she needed to keep her job, so, in effect, her writing did shift away from feminism.

In 2000, Keller published *The Century of the Gene*, which coincided with the first published results from the Human Genome Project. In the book, she questioned the concept of the gene, seeing its function as more interactive with its environment than had been previously acknowledged. For example, she pointed out that DNA could not replicate itself unless the necessary enzyme or protein were present. She questioned the concepts of the central dogma and the master molecule as promoted by Watson and Crick. She also questioned the metaphor of DNA as the "book of life" and confirmed that the book was much more complicated than had been assumed in the 1960s. "Genes do not simply *act*," she wrote. "They must be *activated*." New findings in molecular genetics, she argued, threw the very concept of the gene as a functional unit into question. "Perhaps," she suggested, "it's time we invented some new words."[16]

Keller was proud that this book was not about politics or power. Nor was it a feminist text. In her view, it was purely scientific reasoning based on her knowledge of genetics, and she hoped that scientists would be her main audience. However, they largely ignored her. It felt like the final straw. Keller was weary of being rebuffed; the 1990s had been a series of painful events, and she believed they were affecting her health. In 2001, when she was sixty-five, she suffered a heart attack and went into cardiac arrest. After she recovered, she decided to reduce her teaching load to half-time, though she did continue to publish and give talks. In 2006, she retired from MIT, still hoping that other scientists would applaud her work. But there was something else going on at MIT, something that would soon make news around the world.

CHAPTER 19

THE MIT REPORT AND THE STRENGTH OF INSTITUTIONAL POWER

Although Evelynn Hammonds and Helen Longino didn't mention Nancy Hopkins in their 1990 essay "Conflict and Tensions in the Feminist Study of Gender and Science," they might as well have. Hopkins, who had studied molecular biology with James Watson, was in exactly the position that they were describing. She was not interested in feminist science studies at all. In fact, she proactively avoided it. In 1976, she had written in the *Radcliffe Quarterly* about how difficult it was for a woman to be both successful in science and successful as a wife and mother. After her divorce, she'd decided that she would not remarry or have children. She was choosing science and, in doing so, thought that she'd solved her concerns about being a woman in science. She was wrong.

For fifteen years, between 1973 and 1988, Hopkins and her students at MIT studied tumor viruses in mice. They published more than forty articles, including an important one in the *Journal of Virology* explaining that one protein determined the host range, or the type of cell, in which certain leukemia viruses can grow. Yet despite her contributions, she struggled for recognition, particularly within her own institution. Hopkins wondered if this was because she was working on medical research, a field that was dominated by men.[1]

Meanwhile, as Hopkins was thinking through this idea, she sat down for a 1988 interview with Bonnie Spanier. During the conversation, she expressed gratitude for Jim Watson's role as a supportive mentor and argued that such support was essential for success in science. She said that her years as an assistant professor at MIT had been difficult and that the pressure on women was high, especially if they did not yet have tenure. She recalled the sense that something was going on in her department that she couldn't pinpoint, a feeling that culminated in the moment when the chair told her that she would not be considered for tenure because a powerful colleague did not like her. Although Hopkins did achieve tenure eventually, the process was excruciating. Yet, having survived, she continued to find her life in the lab fascinating. And she still didn't look at exclusion in the workplace as a factor in her struggles.[2]

For a time, a Black postdoctoral fellow worked for Hopkins in her lab. He was smart and dedicated to his work; but during his first year in Boston, he'd had to move several times because he'd been harassed out of one home after another. In the interview with Spanier, Hopkins recalled that it eventually became impossible for the man to focus on his high-powered research. On more than one occasion, after he'd tried to make a purchase at a department store, she had received phone calls asking if he really worked at MIT. She had never received such calls about anyone else. "That's why there aren't any Black people at MIT," she told Spanier, "and that's why I believe there are fewer women. If you make it that much harder, the person just simply has to be a genius." Watching this man's experience gave Hopkins some insight into discrimination; she could see that he was fighting impossible odds. It also gave her a sense of how women could be discriminated against. Yet "I was a privileged person. I went to a private school. I had a nice family. Jim Watson liked me. I couldn't imagine that I was the object of discrimination." Hopkins didn't want anyone to see her as inferior. She felt that this would be shameful, so she continued to deny that discrimination had anything to do with her.[3]

Unlike Banu Subramaniam and Anne Fausto-Sterling, Hopkins did not gravitate toward feminist critiques of science as a way to continue with her lab research. In fact, she did not know either of these women or their work. She found it difficult to simultaneously retain her love of science and her awareness of discrimination. She felt that, if she paid attention to the discrimination, she wouldn't be able to continue with her science.[4] She still believed in a meritocracy, for white people.[5]

Then, in 1987, Ruth Perry, the founder of the women's studies program at MIT, convinced Hopkins to co-teach a course about genetics with Caroline Whitbeck, whom Hopkins described as a "radical feminist philosopher." Whitbeck was an expert in ethics and engineering who had attended Peggy McIntosh's seminar at Wellesley on women and science. Hopkins found the course one of the most interesting she had ever taught, but the two women argued vigorously all the time. Whitbeck "totally opened my mind to how gender was a social construct. I was blown away," said Hopkins. She avoided talking to her science friends about the course, but it was an important step in her own education.[6]

In the early 1990s, after years spent studying and publishing about tumor viruses, Hopkins started to think about changing her area of research. At first, she considered shifting to AIDS; but the field was dominated by a small group of aggressive men, and by this time she was aware that not all male scientists viewed women as full and equal colleagues.[7] Hopkins was interested in genetics and human behavior. She could keep working with mice, making a mutation in a gene to see if it affected the behavior of the mouse. But she was tired of mice. She considered working with human genetics, doing research on large families, but this would require a very large lab and lots of money and, again, the area was dominated by men.

Hopkins had always seen sociobiology as common sense—of course, behavior was affected by biology—but she had never thought of it as her kind of science. For her, real science was laboratory-based. Now the Human Genome Project was exciting. She decided she wanted to find out how genes affect behavior by going into the lab and looking at the mechanics.[8]

One evening, while chatting at a cocktail party at Cold Spring Harbor, Hopkins learned that a German researcher, Christiane Nusslein-Volhard, known as Janni, had begun genetics research on zebrafish. Hopkins was intrigued. Nusslein-Volhard had previously worked on identifying the genes necessary for development in the fruit fly, *Drosophila*, which had eventually won her the Nobel Prize. This was the same field of developmental genetics that Fausto-Sterling had worked in during the 1970s.[9]

Currently, Nusslein-Volhard was working to find out what genes were necessary for the embryo of a zebrafish to develop. Hopkins was excited, so much so that she went to Germany for a sabbatical, hoping to learn from Nusslein-Volhard. "My love for science returned because working with Janni was so much fun. It was a joy to be back in the lab."[10]

Nusslein-Volhard would hand Hopkins a plate of cells. "Watch this under a microscope," she said. As Hopkins watched, each cell divided every twenty minutes right in front of her eyes. These single-celled fertilized eggs turned into little embryos. By the next morning, their little tails were wiggling. Hopkins found the process breathtaking.[11]

Unlike mouse embryos, which are inaccessible inside their mother's womb until birth, zebrafish develop outside the mother and are completely transparent for several days so that researchers can see their growth. Any mutations in the developing fish are visible as well. Hopkins decided that she would not focus on the genetics of behavior in zebrafish; the technology to do so was not yet available. Instead, she would use zebrafish as a model to study developmental genetics—that is, the genetics of a developing embryo.

Hopkins returned to MIT with twenty-three fish, and she immediately ran into obstacles. She needed more space for fish tanks, and the administrators in her department wouldn't help. It was clear that she had less space to work in than her male colleagues did. At the time, she was in her late forties; her age and status should have made it easy to request more space, and she wasn't asking for much, especially when compared to others in her department.[12]

One night, Hopkins got down on her hands and knees with a tape measure and measured the space that her male colleagues had and compared it with her own. Maybe if she could provide the exact measurements she could convince the man in charge of allocating space that she needed an additional two hundred square feet for the tanks. Hopkins spoke to the dean about the matter. He listened but didn't take immediate action. After months of struggling, she then went to the provost, who sent her back to the dean, assuring her he would respond. Hopkins was impatient to move forward with her research, but she was also seeing a bigger problem. And she found it humiliating to have to ask these powerful men for help.[13]

Hopkins knew that science was supposed to be a meritocracy. Yet she had started to notice that, when a man made a scientific discovery and a woman made one of equal importance, the man and his work were valued more highly. She was seeing that women were often invisible and that their work was often attributed to someone else.[14] "At this point, it dawned on me that this strange truth I had discovered might be the most important scientific discovery I had ever made. In fact, it was so important it deserved a Nobel Prize." Hopkins was talking about the psychological concept of cognitive unconscious bias, also called implicit bias. At the time, she thought she was the only person to have discovered this, and she decided that she had better tell the university so that they could fix the problem.[15]

Back in 1976, Hopkins had thought that the only obstacle for women in science was having children and working long hours in the lab. Somehow, she had put the issue of gender bias and discrimination aside to focus on her work in the lab. She had avoided feminists, seeing them as troublemakers. She said Ruth Hubbard and Evelyn Fox Keller scared her. But in the mid-1990s, she saw things differently: "It took such a surprising form that it took me twenty years to recognize it."[16]

Hopkins wrote a letter to Charles Vest, the president of MIT. "There's a terrible chronic problem in this institution," she told him. "I'm sure you don't know about it, because if you did, you'd certainly want to fix it." She showed the letter to a male friend, who said, "You can't send that letter to the president of MIT." So she decided to check with a woman biologist on campus, Mary Lou Pardue. She didn't want to offend Chuck Vest.

Hopkins didn't know Pardue very well but approached her anyway. Over a lunchtime sandwich in a noisy café, she pulled out the letter and asked Pardue to read it. Nervously, she watched her face for a reaction, but there was none. Hopkins wondered if she had made a mistake.

When Pardue finished reading, she looked up and said it was a great letter and that she wanted to sign it too. Pardue said she'd long thought that tenured women faculty at MIT were not treated equally. Hopkins was shocked. She couldn't believe she'd found another person who felt the same way. They looked at each other, and Hopkins asked, "You don't suppose there could be others who agree with us?" The two decided to approach Lisa Steiner, who, back in 1967, had been the first woman ever hired in biology at MIT.[17]

Steiner did agree with them, and now the three scientists began to collect data about the situation. They planned to conduct an informal poll of all of the tenured women faculty members in the six departments of the School of Science. They made a list of names and discovered that, thirty years after passage of the Civil Rights Act, there were still only fifteen tenured women in the School of Science as compared to 197 tenured men. In other words, only 8 percent of the faculty were women.[18] So, to enlarge their sample, the three expanded their list to include two women faculty members in the School of Engineering who had joint appointments in science.

After studying the poll responses as well as MIT's internal institutional data, Hopkins learned that the percentage of women faculty in science at

MIT had not changed since the 1970s, when Hopkins had been hired. The women she had polled could see the discrimination but were afraid to say so. As Hopkins explained, "in a meritocracy, if you say you're discriminated against, people will think you aren't good enough." Now, however, sixteen tenured women scientists were willing to raise the issue. Of this number, four had won the U.S. National Medal of Science, and eleven were members of the National Academies of Science. All had had long and distinguished careers. Nonetheless, they had always, to this point, been reluctant to discuss gender bias. Even now, the topic felt embarrassing and awkward, and they wanted to raise their concerns quietly.[19]

There had never been a woman department chair in the School of Science. In fact, only one of the women had ever been involved in administration. So the group wasn't sure what action to take first. Eventually, they decided to approach the dean of science, Robert Birgeneau. The dean was receptive and authorized the formation of a committee that included some of the sixteen women plus three men who held powerful positions and supported the process. The women involved came to realize that their power lay in the cohesion of the group, rather than as individuals with grievances, and that they also had power as tenured faculty. Thus, for now, the group didn't reach out to younger, untenured women on the science faculty, for fear of jeopardizing their careers. Instead, committee members spent two years systematically gathering data and analysis on women faculty members' experiences and challenges. And in the summer of 1996 they produced a 150-page internal report showing that women faculty at MIT had less lab space and less pay.[20]

The women did not want to be seen as complainers, so they operated quietly. Because they were focused on ensuring equity for women faculty in STEM fields, they did not pull in support from schools outside of science or engineering. This meant that there was no question of inviting Evelyn Fox Keller or Evelynn Hammonds to join their efforts—though Hopkins did privately seek out the assistance of Ruth Perry, a feminist activist and the founder and head of MIT's women's studies program.[21] Notably, all of the sixteen women scientists were white. Years later, when asked if the group had considered issues around race, diversity, or inclusion at MIT, Hopkins simply said no.[22]

In a 2008 interview with Hopkins, John Hockenberry remarked that this story of inequity at MIT sounded like it could have taken place in the 1970s or 1980s. He found it was shocking to learn that it had happened in the 1990s. Hopkins replied:

> Looking back on it, I think I see what happened. I think that in the 1970s, civil rights, affirmative action opened the doors to the universities. A tiny number of women came in. And people thought, oh, well, that was easy. Those women began to progress through the system and, like me, they all felt that civil rights and affirmative action had solved the problem by opening the doors.[23]

She thought it had taken them twenty years to figure out the truth because they were all fixated on their science and believed that science was completely based on merit. Over time, however, this conviction had begun to falter. The women had started to wonder if people could make equal discoveries and not be valued equally. As Hopkins and her colleagues came to realize, this was exactly what was happening. Yet they were not sure what to do about it. They did not want to be political. They did not want to be radical. So they kept working at their science, even as they saw the problems and the discrimination, even as they felt marginalized and isolated, even as they were paid less and given fewer resources for their work.[24]

As Hammonds had written in 1990, it's hard for women scientists to take on these issues.[25] Moreover, feminist critics of science during the 1970s and 1980s and women working in feminist science studies in the 1990s had not reached out successfully to many women scientists.

After the group quietly raised their concerns with Dean Birgeneau, he made some changes that addressed the inequities of individual women on the science faculty. As a result, Hopkins was able to expand her lab and bring in the fish tanks. At the height of her successful research, she had 4,000 fish tanks and between 50,000 and 100,000 fish in the lab at any one time. She brought on board a team of twenty-five people who screened close to a million fish, examining them one by one to learn which had defects as a result of gene mutations. She was happy.[26]

Most of the women who had joined Hopkins to write the internal report were also happy. They went back to their labs and their classrooms, pleased that some level of fairness had returned. Almost no one else at MIT knew what had happened.

But then, in 1999, the chair of the faculty, Lotte Baylin, a social psychologist, asked that a summary of the report be published in the faculty newsletter.

Before it went out, she asked President Vest if he wanted to write an introductory note; neither she nor Hopkins wanted readers to think that MIT had been blindsided. Vest agreed. In his note, he wrote, "I have always believed that contemporary gender discrimination within universities is part reality and part perception. True, but I now understand that reality is by far the greater part of the balance." Hopkins later recalled her deep emotion on reading that note for the first time. Vest had acknowledged publicly that discrimination was a problem and that MIT had to do something about it.[27]

As soon as the newsletter was released, media around the country jumped on the story, including the *New York Times*, which published a front-page article about the report. Hopkins had not anticipated that it would be a big story; but from the moment the report went out, her phone never stopped ringing. She remembered sitting in her office, picking up her phone, and suddenly hearing a voice say, "Hello, you're on the air in Australia."[28] Hopkins found the reaction overwhelming. "I didn't understand the breadth of this problem," she later said. "I'm embarrassed to say it. Where was I? I think I spent too much time in the lab. I mean, I just didn't get it."[29]

For months, she was inundated with emails from women around the United States and beyond, all of them saying, "This is my story too." She received more than 1,000 invitations to talk about women in science. She accepted more than 150 and spoke at universities and companies across the country. Within weeks, President Bill Clinton and First Lady Hillary Rodham Clinton had invited her to an event at the White House.

Evelynn Hammonds later pointed out that this was not the first report to document discrimination and animosity toward women scientists. They had produced many similar reports over the past hundred years. In her view, the 1999 MIT report was different only because a man in power, President Charles Vest, "finally read it and said, we have to do something about it."[30] The report's global reverberations illustrated the impact of institutional power. Just as a collective has more impact than an individual, a powerful institution has a greater impact than a collective. The broad influence of the MIT report also illustrated the power of elite universities. If a less prestigious institution had released the report, it would have had less force.

In 2000, at a meeting sponsored by the National Research Council on women in the chemical sciences, Hopkins talked about the experience of writing the MIT report. Earlier in the day, a high-level government official had told attendees that the economic future of the United States depended on increasing the number of women and minorities in science and engineering. "This seems so odd," said Hopkins, "given the fact we have spent two days talking about how women are being driven out!" She argued that it was necessary to "call a moratorium" on further studies to document gender bias. As she explained to the audience, the problem was so common that it was necessary to institutionalize the actions needed to guard against it, correct its consequences, and raise consciousness to eliminate it.[31]

At question time, Suzanne Franks from Kansas State University stood up and asked Hopkins about the difference between what Franks called "the liberal project in science" to increase the numbers of women and the more "radical transformation project" to change the structure of science. Franks's view was that science needed not just women scientists but feminist scientists: "Do you have thoughts on what people at this workshop can do to make it possible to be feminist scientists?"[32]

"I think I'm the wrong person to answer that question," replied Hopkins. People at MIT had bought into the system, she explained. They weren't trying to change it; they just wanted to be treated fairly within it. She pointed out that Lotte Baylin, as chair of the faculty, was focused on changing the structure of the workplace and that it was important for young women to say how this should be done.

Soon after the MIT report was released, Evelynn Hammonds got a phone call from President Vest. At the time, she was away from MIT, serving as a visiting professor at the University of California, Los Angeles. Vest asked how she was doing and when she was coming back.

"I'm not sure," said Hammonds, "it's nice out here."

"This report on women at MIT didn't really address issues of race," said Vest.

"I know," said Hammonds.

Vest mentioned that Wesley Harris, a Black aeronautical engineer at MIT, was currently writing a report about issues of race. But it wasn't really dealing with gender.

"I've noticed that," said Hammonds.

"I really need you to come back and help us with this," said Vest.

Hammonds had been thinking of staying at UCLA, but she found it difficult to turn down Vest's request. He wanted her in a position of leadership and offered her a position as secretary of the faculty. She didn't find it an especially interesting job, but she did appreciate the ability to be part of the leadership team and to meet regularly with faculty, the president, the provost, and the chancellor. Hammonds knew that it was important for everyone to address issues of exclusion at MIT in terms of both gender and race. Now she was in a position to influence MIT leadership includinging Lotte Baylin and Nancy Hopkins.[33]

In later reflections on the 1999 MIT report, Hammonds looked back at Margaret Rossiter's first volume of *Women Scientists in America*. "Few people," she said, "want to believe that science could be riddled with [the] . . . seemingly immaterial prejudices that Rossiter's story chronicles." Yet those prejudices accumulate into inequity. Hammonds pointed out that many people, Hopkins included, thought that the Title IX and the Civil Rights Act would automatically topple barriers. But when legislation is not enforced, there is no change.[34]

Long afterward, Hopkins reflected that it was Hammonds who had taught her about the double bind, a crucial key to her understanding of the circumstances facing women of color in the academy. Hopkins had not seen how much easier it was for white women in a white, male-dominated institution. Unlike Hopkins, Hammonds could not afford to be reluctant in her feminism.[35]

Anne Fausto-Sterling later said that the MIT report did not reveal anything she didn't already know. Maybe it had caught the attention of more mainstream women scientists at Brown, but to her it was nothing new. She commented, "Nancy Hopkins was at an elite institution, worrying about elite women, and that's fine, but it's limited."[36] Unlike the work of feminists over time, this report was limited to the challenges of individual women and was not part of a larger social movement. Yet it had struck a chord with mainstream women scientists in a way that feminist science studies never had.

In 2000, Hammonds was named the founding director of the MIT Center for the Study of Diversity in Science, Technology, and Medicine. Finally, she was in a position to address issues that she had been thinking about for the past twenty-five years. In January 2002, she hosted a two-day conference for women-of-color faculty in science and engineering. They arrived from around the country to discuss their continued career barriers inside the academy. Hammonds modeled the conference on the meeting that had taken place more than twenty-five years earlier, the one that had produced "The Double Bind." The problems that Shirley Malcom and her co-authors had described still existed: there were only four tenured women of color at MIT, out of a full-time faculty of about 950 professors.

Nancy Hopkins attended the conference. "We may look back on this meeting as a critical turning point," she said, commenting on its role in advancing the participation of MIT faculty of color in science and engineering. One goal of the gathering was to explore institutional arrangements that had helped or hindered their work in the academy. Hopkins found the data "eye opening," revealing that "we as faculty have failed our students, our institution and most of all our nation."[37]

During her speaking tour, after the release of the MIT report, Hopkins had repeatedly asked, "Why would anyone think that a woman's identical work is less good than a man's?" If we examine the important work of women scientists during the past 150 years, we see that the answer is clearly related to social discrimination and the new barriers that are constantly set up against women scientists.[38] But there are always powerful people willing to come up with other reasons: that women aren't good enough, that men and women of color aren't intelligent enough. Nancy Hopkins and Evelynn Hammonds had no idea that, within five years of the MIT report, the president of Harvard University would suggest this same tired answer again.

CHAPTER 20

REJECTING "WARMED-OVER SOCIOBIOLOGY"

In January 2001, the presidents of Harvard, Yale, MIT, Princeton, Stanford, the California Institute of Technology, the University of Pennsylvania, the University of Michigan, and the University of California at Berkeley came together for the first time to talk about women in science. The consortium, which would become annual, was a direct result of the 1999 MIT report and was hosted by MIT.

Shirley Malcom of the AAAS welcomed the participants. Twenty-five years after co-authoring "The Double Bind," she was still working to open the sciences to those who had historically been excluded. Her opening remarks emphasized the need for both stories and statistics. President Charles Vest of MIT agreed, arguing that it is necessary to learn from both data and individual women's experiences. Ironically, however, at this meeting of forty-two people talking about women in science, there were only three women of color in the room: Malcom; Ruby Lee, a professor of electrical engineering at Princeton University; and Gertrude Fraser of the Ford Foundation.[1]

For several years, the consortium met annually. But when Larry Summers was named the new president of Harvard, he did not make it a priority to attend. In fact, he did not think that hiring women should be a priority for the university. Several articles in the *Boston Globe* noted that the numbers of women being offered tenure at Harvard had fallen substantially under his leadership. When Nancy Hopkins brought up the issue of women in science during a meeting, Summers made it clear that he wasn't particularly interested.[2]

Ironically, it was during Summers's tenure that Evelynn Hammonds decided to leave MIT and take a position at Harvard. She had spent nine years at MIT; and in her most recent position as founding director of the Center for the Study of Diversity in Science, Technology, and Medicine, she had built a well-funded hub for scholarship and conversation. Yet despite President Vest's backing, many people in the Department of Science, Technology, and Society were not supportive of the center. And because it did not sit within a particular department, Hammonds and her team were isolated.[3]

Hammonds considered moving to Brown. But then an opportunity came up at Harvard: a joint position in the history of science department and the African American studies program. Hammonds had no illusions about what it would be like to work at Harvard. Yet few, if any, of the nation's African American studies programs were focused on science and medicine, despite the fact that Black people had made significant contributions in those fields. Likewise, few history of science departments were seriously addressing race and racism, and she liked the idea of bringing these two areas of interest into one job. Another appeal was the thought of working with African American students. At MIT, she'd had few such opportunities.[4]

In October 2003, Hammonds became the fourth African American woman to achieve tenure in the faculty of arts and sciences at Harvard.[5] It had been a momentous year for her. In January, she and her partner, Alexandra Shields, had welcomed a son into their family, and soon their family would strengthen in another significant way. As a result of a Massachusetts court ruling, the pair would become one of the first same-sex couples in the country to receive a marriage license. On May 17, 2004, Hammonds and Shields were married at Cambridge City Hall just after midnight. Afterward, they walked out the front door into a sea of people celebrating on Massachusetts Avenue. Among them were Anne Fausto-Sterling and her partner, Paula Vogel, who had driven up from Rhode Island to get married and join the festivities.[6] Although state legislators and the governor of Massachusetts, Mitt Romney, had tried to overturn the court ruling, they had failed, and the crowd was jubilant.[7] Some people were singing; others were waving glow sticks and balloons in front of TV cameras. Hammonds was elated.[8]

———

In January 2005, Nancy Hopkins was scheduled to give a talk on women in science at a conference hosted by the National Bureau of Economic Research. The event would take place in Cambridge, and the only reason she agreed to speak was because she knew that women faculty from Harvard were planning to attend. Several had told her that they were struggling to convince President Summers about the serious problems facing women at Harvard, and he was not listening to them.[9]

The topic of the conference was how to increase the representation of women and people of color in the scientific workforce and STEM, and Summers was scheduled to speak at a lunchtime session. Hopkins's talk had taken place earlier that morning, and he did not arrive in time to hear her speak.[10] But she planned to be in the audience for his.

After grabbing a sandwich at the buffet table, Hopkins sat down at a long seminar table, waiting for Summers to arrive. More than fifty university professors and administrators were packed into the conference room, including Shirley Malcom, who was sitting next to Hopkins. Everyone was waiting to hear what the president would have to say. Hopkins knew that, given the power of his position, he could quickly affect global attitudes about the issue. But would he make the effort?[11]

When Summers arrived, he walked directly to the podium. "I've made an effort to think in a very serious way [about this problem]," he said. He noted that the dearth of women in science was not the only example of a group that was significantly underrepresented in an important activity. For instance, he said, there were also few Catholics in investment banking, few white men in the National Basketball Association, and few Jews in farming and agriculture.[12]

In his opinion, three reasons explained the underrepresentation of women faculty in science, technology, and engineering at Harvard. He presented these reasons in what he saw as their order of influence. The first, and most significant, was the difficulty of juggling work and family. He called this "the high-powered job hypothesis," and Hopkins agreed with him on this point. The second was a possible biological and genetic difference between men and women, which he called "a different availability of aptitude." The third was linked to issues of socialization, patterns of discrimination, and bias, all of which, he declared, were much less important than most people thought. Economic theory, he said, would drive out bias. (Summers himself was an economist.)[13]

Hopkins, who had been shuffling papers, thought that perhaps she hadn't heard him correctly. But, in fact, he'd meant exactly what he'd said. In his discussion of the differences in aptitude between men and women in the sciences, he stated that some people are several standard deviations above the mean in scientific aptitude and that, at the high end of the curve, the male-to-female ratio was five to one. He acknowledged that some of the papers presented at this conference pointed out that ability tests "are not a very good measure and are not highly predictive with respect to people's ability." Yet even though he admitted that it was possible to argue about the data, "there are some systematic differences in variability in different populations."[14]

Hopkins couldn't believe that Summers was choosing to highlight aptitude over discrimination as a reason for the absence of women in science. She had spent the past decade listening to women scientists who had suffered from discrimination, and Summers was dismissing it as an issue. Hopkins turned to Malcom and asked, "Do you think we should leave?" Malcom said she wanted to but couldn't because she was the next scheduled speaker. Hopkins, however, felt ill. She closed her laptop, stood up, and walked out.[15]

After Summers finished speaking, Malcom stood up. Composed and direct, she said, "It is wrong-headed to dismiss biology," but also to give it too much weight. If people have "different opportunities for socialization, there is good evidence to indicate that it w[ill] have ... different outcomes."

Then Denice Denton, the chancellor of the University of California at Santa Cruz spoke up. "In the spirit of speaking truth to power ... a lot of us would disagree with your hypotheses and your premises," she said.

Summers responded, "Fair enough."[16]

One can trace a straight line between Summers's remarks in 2005 and E. O. Wilson's claims in *Sociobiology* in 1975. Summers had based his arguments on a chapter on gender in Steven Pinker's *The Blank Slate* (2002).[17] Pinker is a scholar of evolutionary psychology, which centers around the idea that much of human behavior has roots in evolutionary theory. When sociobiology in relation to human behavior became controversial in the 1970s, its scholars shifted away from that name and began to describe their field as evolutionary psychology. By the 1990s the discipline was thriving, and its popular literature emphasized that important sex differences resulted from evolutionary pressures—that is, from biology, not socialization.

Soon after *The Blank Slate* was published, Pinker accepted an invitation from Summers to leave his post at MIT and take up a position at Harvard. He was already a popular author by the time the book appeared, and he took advantage of his popularity to make inflammatory claims—for instance, applauding "equity feminism" and deriding "gender feminism" as promoted by people such as Anne Fausto-Sterling. In his chapter on gender, he wrote extensively about women scientists and the so-called "leaky pipeline." While he did not mention the 1999 MIT report specifically, he did refer to the subsequent 2001 meeting of university presidents. There's "something odd," he said, "in these stories about negative messages, hidden barriers and gender prejudices." He claimed that none of the people discussing this issue ever mentioned an alternative theory—by which he meant innate or genetic differences or intrinsic aptitude. He also argued that, "in a free and unprejudiced labor market, people will be hired and paid according to the match between their traits and the demands of the job." Clearly, Pinker had not read any of the large body of literature on the history of women in science.[18]

On the day of Summers's speech, Marcella Bombardieri, a journalist at the *Boston Globe*, sent Hopkins an email. Though she was following up with Hopkins about another project, she asked in passing how the meeting had gone. Hopkins said that she had walked out. Intrigued, Bombardieri contacted numerous other attendees, and on the following Monday she broke the story that Summers had "sparked an uproar."[19]

After the *Globe* story, a storm of media coverage appeared around the world. While plenty of voices criticized Summers, Stephen Pinker led the charge in defending him, and conservative pundits followed suit. In one article, the nationally syndicated columnist George Will called Hopkins a "hysteric."[20]

Not everyone was focusing on the scientific debate. Many were more concerned about the impact of Summers's comments on policy and hiring at Harvard. In the 2002–3 academic year, four out of thirty-two tenure offers had gone to women. In 2003–4, that number had dropped to only one of twenty-two tenure appointments.[21]

Evelynn Hammonds had been one of Summers's 2002–3 appointees, and she was appalled by his remarks at the conference. "I couldn't believe he said that." In an interview, she made it clear that the best current social scientific literature was proving that innate differences between men and women didn't account for differences in women's achievements in science and engineering. She was shocked that Summers was willing to discount discrimination and the many serious barriers to women in the sciences and return to innate sex differences, which had been evoked for centuries to suggest that women were inferior. "I was disappointed in him for taking that stand, which I believe was fundamentally misinformed/uninformed."[22]

At the time, Hammonds was serving on the faculty's Committee on the Status of Women. Four days after Summers's talk, its members published an open letter stating that his remarks "serve[d] to reinforce an institutional culture at Harvard that erects numerous barriers to improving the representation of women on the faculty and to impede our current efforts to recruit top women scholars." The letter also expressed concern about the impact of his comments on young science students: "The biggest problem with female science students is confidence. . . . When they are sitting there constantly saying, 'Am I smart enough? Am I smart enough?' it doesn't really help when the president of the university says, 'Maybe you're not.'"[23]

Summers was not familiar with the history of exclusion that Roger Arliner Young, Ernest Just, Ruth Hubbard, Rita Arditti, Evelyn Fox Keller, Evelynn Hammonds, and many others had experienced in science during the past century. He had not read the growing literature in feminist science studies. He had not read Margaret Rossiter's writings. He had simply read Steven Pinker's *The Blank Slate.*

Summers was widely criticized in the media throughout the country and internationally.[24] Yet, the *Harvard Crimson's* coverage of the incident showed that the campus response was markedly different from how students had engaged with sociobiology thirty years earlier. Most students in the early 2000s were not activists, and there was no broad protest of the kind that had occurred in the days of the Sociobiology Study Group, Science for the People, Students for a Democratic Society, and the Committee against Racism.[25] Without a unified organization promoting feminist science studies, the response to Summers's remarks came largely from individuals.

In a Harvard history of science thesis, one undergraduate, Hana Rachel Alberts, posited that people in the 1970s had tended to look at biological determinism and conflate concerns about racism and sexism. Alberts suggested that the protests around sociobiology were primarily attempts to point out its racism. In contrast, the response around sexism had been muted and was almost written out of history. As the field of evolutionary psychology had grown, it had focused predominantly on sex differences rather than race. Thus, in 2005, the academy seemed willing to accept biological explanations for certain sex differences, just not as the reason for the numbers of women in science.[26]

Hammonds did not view the muted response at Harvard as ideological. "I read it as a profound lack of knowledge," she said. It also reflected a change in course offerings. After Ruth Hubbard retired in 1990, her class "Biology and Women's Issues" was no longer offered, and no comparable class had since appeared in the undergraduate curriculum.[27]

In 2005, when Summers made his remarks, Hubbard was eighty-one years old and frail. But that didn't mean she wasn't listening. In an interview, Alberts asked why she hadn't protested the president's claims. Hubbard retorted, "It's because we feel it's warmed-over sociobiology, and because it's kind of boring to say the same things over and over and over again."[28]

At a Radcliffe seminar in March, Hopkins talked about why she had walked out of Summers's lecture. The history of discrimination against women in academia was extensive, she said, and statements such as his could justify continued discrimination. During the question-and-answer period, Hopkins declared that, when a biologist finds a difference in men's and women's brains and concludes that this difference explains everything, the biologist is not doing science.[29]

In April, Evelyn Fox Keller spoke at another Radcliffe event, where she chronicled the history of the nature-versus-nurture debate from Victorian England to the present, demonstrating how a "most peculiar equation has come to prevail." Critics of heritability measurements were now seen as political, whereas the defenders of heritability were seen as scientific. Both assumptions were incorrect, and Keller argued that research must continue. More data were needed, not more debate.[30]

Yet, as Fausto-Sterling had already written, more data does not always clarify the science. Hubbard also disagreed with Keller. To her, the issue wasn't about the data. It was about power. She said, "If Steven Pinker says something, people can take it or leave it. If Richard Lewontin or I say something, you can agree or disagree." But when the president of Harvard said something, the situation was different. He had the power to make decisions about hiring and to influence students' lives. His statements deserved a different level of attention.[31]

The changing views around these long-discussed topics reflected media coverage of the Human Genome Project (which President Bill Clinton had announced with great fanfare in 2000) and the rapidly expanding field of genetics, which had influenced public opinion about many biological issues. Countless newspaper articles were discussing newly discovered genes that linked to one human trait or another. The public was enthusiastic about the potential of genes to explain various diseases and behaviors. But few people realized that biology is not fixed and that the presence of a gene does not confirm a certain outcome. Rather, genes interact with and are influenced by their environment.[32]

For the moment, the pendulum was swinging toward genetic explanations. However, by 2010, it would eventually swing back toward greater skepticism about the role of genes in every human trait. Members of the scientific community would become more circumspect about the impact of the Human Genome Project, and this disappointment would lead to growing critiques of the role of biology and genetics in human behavior. The interactionist model would gain traction, and the nature-versus-nurture debate would begin to unravel.[33]

Anne Fausto-Sterling was at home in Rhode Island when she heard about Larry Summers's remarks at the conference. She was irritated, but she vowed that she wouldn't get sidetracked by them. She wanted to continue focusing on her own research, her own writing. Banu Subramaniam felt the same way. Why let his ignorant comments distract her?[34]

In the end, the debate over Summers's talk became the basis for a general critique of his leadership style. Eventually, after receiving a no-confidence vote from the Harvard faculty, he resigned as president in February 2006. By June he was gone from the campus.[35]

Years later, Hammonds reflected more broadly on responses to Summers's 2005 remarks. She pointed out that men had been saying such things since Darwin. What was new was the global rejection of what Summers had said, which was the real indication that attitudes had changed.[36]

Also, in relation to innate sex differences as an explanation for the smaller number of women in science, "history tells us that this was never really the issue," Hammonds said. "It was *made* the issue to explain away the practice of exclusion."[37]

Banu Subramanian would lament the lack of an organized institutional response to Summers from feminist scientists. Yet, a published conversation between Subramaniam and Hammonds *had* shed light on the strengths and challenges in the field of feminist science studies.[38]

CHAPTER 21

REFLECTING ON A FIELD IN MOTION

While working on her dissertation, Banu Subramaniam had loved formulating hypotheses, designing experiments, being out in the field, planting seedlings, watching them grow, entering her data into the computer, and watching patterns emerge. So, in 2001, after accepting a position as assistant professor in the women's studies department at the University of Massachusetts Amherst, she told herself she wanted to continue that experimental work.

Feminist science scholars had influenced Subramaniam's studies, and she looked for ways to incorporate these themes into her science research. Evelyn Fox Keller's writing on language and science was compelling. Anne Fausto-Sterling's work encouraged her to look closely at how her experiments were designed. Helen Longino and Sandra Harding pushed her to think hard about the conclusions she drew and how she had come to them. Ruth Hubbard and Evelynn Hammonds had shaped her thinking. And now that she was in western Massachusetts, a two-hour drive from Cambridge, she was able to start attending Hammonds's Workshop on Race and Science.[1]

In Subramaniam's view, feminist science studies had been using insights from feminist scholarship throughout the 1990s to deconstruct the sciences. Now she hoped to apply these insights in a project of reconstruction and to influence innovative practice *within* the sciences. As Emily Martin and the Biology and Gender Study Group had done with cell biology and fertilization, she wanted to engage in scientific research that was informed by feminist scholarship in science studies.[2]

In 2001, Subramaniam and her co-editors, Maralee Mayberry and Lisa H. Weasel, published *Feminist Science Studies: A New Generation*. In the introduction to this anthology of essays, they acknowledged that it was difficult to define a field in motion and thanked the many feminists who, over the past two decades, had "highlighted how gender, race and class are implicated in the development of science."[3] In her own essay for the book, Subramaniam described her personal transformation, which resembled Anne Fausto-Sterling's when she'd tried to teach her embryology course after a decade spent exploring science studies. Subramaniam wrote that her work going forward would have to be interdisciplinary. It would no longer simply mirror the logic of scientific experiments but would also mirror the context of language, history, culture, and politics. "The 'perfect' mirror of nature I had constructed had, quite simply, cracked!"[4]

Subramaniam highlighted a new word that Donna Haraway had coined, *natureculture*, to suggest that our observation of nature exists within culture, that culture influences nature, and that all humans, including scientists, are immersed in both. She believed, as Haraway and Harding had written in the 1990s, that knowledge would always be situated within a historical, geographical, cultural, gendered context. Looking ahead to the new decade, Subramaniam declared that this interdisciplinary work "is the exciting work ahead of us in feminist science studies."[5]

Yet she and her co-editors knew that feminist science studies continued to be plagued with the problem of exclusion that had long haunted feminist scholarship more broadly. The issues that Audre Lorde had raised about the feminist movement in 1979 were still present twenty years later, especially in science: "While it is commonplace to speak of the importance of race, class, nationality, ethnicity and sexuality, in practice, these variables rarely shape the final analysis presented."[6] The gaps in the feminist science studies anthology were not new in the field. As Patricia Hill Collins wrote, looking at these variables as distinct was problematic. Instead, they had to be studied at their intersections. Clearly, the growing scholarship on intersectionality related not only to history, literature, and law but also to science.[7]

In February 2002, Subramaniam and Hammonds attended a conference together. "Balancing the Equation: Where Are Women and Girls in Science,

Engineering, and Technology?," which took place at Barnard College in New York City. The two were the only women of color on any of the panels. Both were struck by how many times this had happened since they'd first met in 1994, and they talked a lot about their experiences and ideas. After the conference, they continued their conversation via phone and email. They shared that they both had been excited by the developing field of feminist science studies in the 1990s. They both remembered waiting for an explosion of scholarship that would lead the field into the future. They'd thought that it would grow in the same way that fields such as critical legal studies, literary studies, and historical studies had grown. But in 2002 they were still waiting.[8]

In their conversations, Hammonds pointed out that the primary schism in the field was between the feminist critiques of science and the efforts to promote women in science—two very different projects. The field of women in science was growing enormously, with large staffs and budgets at universities and scientific institutions, especially since the publication of the MIT report. Yet these efforts were separate from feminist science studies, which were relatively underfunded. Keller had made a similar point in the early 1980s. Yet, if anything, the schism between efforts to promote women in science and efforts to understand science with a feminist awareness was widening.[9]

Hammonds and Subramaniam felt their conversation was capturing something important, and they decided to formalize it as an article. The result was "A Conversation on Feminist Science Studies," published in *Signs* in 2003. By choosing to publish in a women's studies journal, they knew they were talking to other feminists rather than to other women scientists, but it was the best option open to them.[10]

In the conversation, Hammonds pointed out that there was still no coherence in feminist science studies that would suggest a unified area of scholarship. The field was defined by the publication of anthologies, the appearance of scientists on panels and conferences, and the citation of one another's work. But it hadn't cohered in the same way that feminist literary studies or feminist legal studies had.[11]

Subramaniam pointed out that, in the twenty years between 1980 and 2000, only rarely had feminist science studies become a site for creating new knowledge in the sciences. In contrast, feminist work had certainly created new knowledge in literary studies and legal studies. She also pointed out that some of the classic essays in feminist science studies were more than twenty years old. "Critiques of biological determinism, sociobiology et cetera are as relevant today as they were twenty years ago. Certainly, evolutionary psychology seems like yet another incarnation of sociobiology," she said.[12]

"Do you see the invisibility or marginalization of race?" Subramaniam asked Hammonds.

"Yes, I do," Hammonds replied, saying that "the term *woman* had been defined in an explicitly white, Western, and privileged way." While feminists of color had worked hard to challenge that definition, they had not paid much attention to science. The fact that there were more women of color in other fields meant that there was greater analysis of the relationship between race and gender in those fields. An exception was Patricia Hill Collins's 1999 article suggesting that intersectionality could be important for feminist science studies. Hammonds was troubled by the fact that few people writing about gender and science were also writing about race and science.[13]

The marginalization of women of color in feminism and in science compounded the problem. "I can see structures of invisibility just proliferating across feminist studies of science," said Hammonds. "Nationally, programs designed to help women in science have had a difficult time reaching women of color, while programs to help people of color in science have largely focused on men." She said that she had always tried to step into the space. "But by stepping into it, I almost became iconic in that I was in many ways a singular voice." Hammonds noted that, when they were speaking on panels, Subramaniam got figured as "the Indian woman" and she was figured as "the Black woman."[14]

Hammonds looked back on the many years of trying to bring more women of color into the sciences. At first, she said, the focus had been on documenting the absence of women of color and generating recommendations about hiring. However, over time, women of color had not, in general, been fitting into scientific cultures, thriving there, or becoming leaders. As a result, many policy documents were now stating that it was time to look at how these cultures were shaped by the absence of women and minorities and to define the problem as the pipeline.

In the mid-1980s the National Science Foundation and AAAS started to use the pipeline metaphor—that is, promote the idea that students must enter an educational pathway when they are young and continue along it to university and an adult career in science or technology. A few years later scientific institutions began to express concern about a leaky pipeline, a reference to issues of retention.

Neither Hammonds nor Subramaniam liked the pipeline metaphor. In Hammonds's opinion, it told a false story: that there was one way into science and one way out and that a certain set of things needed to happen in the pipe. The implication was that the same things happened there to everybody and the best people fell out at the other end. But things just didn't work that way.[15]

Hammonds was beginning to feel that all discussions about women in science were "trapped in the pipeline," and she pointed out that the metaphor was rarely used about men in science.[16] She believed it was important to ask questions about the role of capitalism, industry, and the state in the funding of science. Subramaniam also criticized the metaphor. "Where is the pipeline going?" she asked. "Does anyone want to go there?" She noted that no one ever asked, Who laid these pipes? "Why are we so invested," she wondered, "in shoving all these young girls and women into the pipeline that is dark and dingy and not very habitable?"[17]

According to Hammonds, in the 1980s and 1990s, there was no unified research program and few graduate students in the field available to take up such questions. To follow developments in feminist science studies, Subramaniam said she had to read journals in other fields: mainstream biology, women's studies, science studies, history of science, philosophy of science, sociology, anthropology, literary studies, women's history, philosophy, cultural studies, ethnic studies, postcolonial studies, and so on: "What was striking was that there was no central journal, no central conference, no central group of scholars that consistently came together, ever."[18] Nonetheless, both Hammonds and Subramaniam had continued to feel that there was something valuable in feminist science studies. The field recognized that science had been a central force in society, and it was trying to understand the relationship between science and gender, race and class, sexuality and colonialism. Yet, as Subramaniam said:

> Until we historicize and situate the practice of science (and why women and people of color don't fit in it) we will never be able to get beyond the myth that these arbitrary codes of "good" scientist are really the best way to produce "good" science. So for me, feminist science studies has a tremendous lot to say to (and learn from) practicing scientists.[19]

Hammonds didn't think the field could develop further without more focused venues for exchange, without a journal, without a conference. But "I wouldn't want people to read our exchange and go away thinking that we or

I am suggesting that feminist science studies should be unified. I believe that diverse perspectives in the disciplines from which scholars in this field come is one of its great strengths."[20]

Subramaniam agreed that with no unified consensus or cohesion, many scientists, such as Nancy Hopkins, would remain unaware of feminist science studies and its importance. Yet, she also believed that the lack of cohesion had resulted in vibrant and diverse contributions to many fields. At the start of the new millennium, Subramaniam and Hammonds hoped that feminist science studies would continue to grow and spread throughout the sciences, influencing labs, permeating classrooms, and shaping our knowledge of the world.

FEMINIST SCIENCE STUDIES
GOING VIRAL?

In January 2021, President Joe Biden appointed Alondra Nelson, an African American sociologist and science studies scholar, to be deputy director of science and society for the White House Office on Science and Technology Policy. In her public statement after the appointment, Nelson spoke of both the Covid-19 pandemic and the climate crisis, saying, "Never before in living memory have the connections between our scientific world and our social world been quite so stark as they are today." Many scientists and scholars celebrated her acknowledgment of the impact of science on society and vice versa.[1]

Over the course of her career, Nelson had relied on Evelynn Hammonds as a mentor, first at the MIT Workshop on Race and Science during the 1990s, then after she accepted academic positions at Yale and Columbia, and eventually when she served as the president and chief executive officer of the Social Science Research Council. The tools of feminist analysis, of Black feminism, and of intersectionality informed Nelson's engagement with science and science studies. They also influenced her scholarship on the societal impacts of new technologies, including genetics, as well as science, medicine, and social inequality.[2]

Nelson's work demonstrated how feminist approaches to science could affect science at the highest levels. Yet some things never change: today,

without intervention from feminist scholars, the media continues to repeatedly misrepresent gender, race, and science. For instance, early during the
Covid pandemic, a series of headlines reported that men were dying of the
coronavirus at twice the rate of women. At first researchers examined innate
biological sex differences to explain the divide, and the media leaped to conclusions. Articles in *The Guardian* and *Nature* suggested that differences in
hormone levels or immune responses between males and females could offer
an explanation. The *New York Times* even suggested that men with the virus
could be treated with estrogen to offer protection.[3]

But this focus on biological sex differences was misplaced and insuffi-
cient. A group of researchers at the feminist, collaborative, interdisciplinary
GenderSci Lab at Harvard, established in 2018, analyzed more than a year's
worth of Covid data. Their results showed that the gender gap was explained
by differences in mask wearing, social distancing, testing rates, and other
such factors—reasons that had nothing to do with hormones or immune
response. As their 2022 paper emphasized, even if sex differences had contributed to health outcomes, researchers needed to explore how those differences interacted with social and cultural factors, an approach to analysis
developed by feminist science studies.[4]

This example also shows that it is still possible to look at gender and science
without considering race and racism. Early during the pandemic, Hammonds
felt a sense of dread. "I knew that pandemics don't produce inequalities," she
said; "they reveal them." She understood that communities of color would be
disproportionately affected by the virus—especially women, who would be
both victims and caretakers. Hoping to influence public awareness, Hammonds spoke to a reporter for *The New Yorker* about how racism was shaping
the pandemic and went on to host a series of seminars at Harvard on the
history of African Americans and epidemics.[5]

In 2022, the science writer Rachel E. Gross published *Vagina Obscura*,
which examined the history of the vast gap in knowledge about women's
bodies. In her years of research for the book, she had found a deep and
continuing disconnect between how many scientists view feminists and the
actual tools that feminist science studies have to offer. "Among mainstream
scientists," she wrote, "the word feminist has often been viewed with disdain,
hostility, and an implicit belief that feminist ideals are incompatible with true
science—that the former is about ideology; the latter, objective authority."[6]

This view echoes the challenges of fifty years ago. When early feminists critiqued the objectivity of science, scientists critiqued feminists as ideological and subjective. Even today, mainstream scientists continue to disdain the tools that Rita Arditti, Ruth Hubbard, Evelyn Fox Keller, Evelynn Hammonds, Anne Fausto-Sterling, Sandra Harding, Donna Haraway, Banu Subramaniam, and many others have constructed over the years. But as Audre Lorde said, the master's tools will never dismantle the master's house. Those tools that feminist science studies have developed are critical to the sciences because they bring in marginalized perspectives, ask new questions, and develop new methodologies that help science correct for gender and racial bias.

After closing down in 1989, Science for the People was revived in 2014 as both an organization and an online magazine. In its reinvigorated role, it now addresses issues such as the climate crisis and Covid-19 as well as new critiques of science in relation to race and gender, releasing special journal issues on topics such as "Racial Capitalism" (2021) and "Beyond Binaries" (2023). The rejuvenation of Science for the People shows that social movements can continue to play a role in the relationship between science and society.[7]

When Donald Trump was elected president in 2016, it became clear that support for science as a national priority was under threat, and concerned scientists and organizations began plans for a 2017 March for Science. Yet even at this crucial moment, organizers were slow to recognize the political nature of science. Stephani Page, a biochemist and biophysicist who in 2014 had created the online community #BLACKandSTEM, helped start a Twitter discussion via the hashtag *#marginsci* that forced planners to reflect on the exclusion of underrepresented groups. The editors of *Science for the People* agreed, writing that any rally against the "attack on science" also needed to address the systemic racism, sexism, ableism, classism, and homophobia of science communities: "These are not simply problems of individual morality but rather are complex structural problems and must be addressed as such through organization and demands beyond the calls for objective or diverse science."[8]

In 2017, Angela Saini, a British science writer, published *Inferior: How Science Got Women Wrong and the New Research That's Rewriting the Story.* In the same year, Jess Wade, a British physicist at Imperial College, began writing hundreds of Wikipedia entries about unrecognized and overlooked women scientists. Wade, who works in experimental solid-state physics, has continued this campaign ever since, working with a growing group of editors, including Anne Fausto-Sterling.

In that same year, the #MeToo movement was highlighting pervasive sexual harassment in Hollywood and other workplaces, including science labs.[9] By 2020, Black Lives Matter was bringing attention to the institutional racism that threads through every aspect of American life, including science, technology, and medicine. As Alondra Nelson said in September of that year, "this is a moment for deeds, not words. To topple the edifice of structural racism that produces 'excess' death in the context of COVID-19 and of life generally will take urgent social, political and economic action, from court rooms to clinics, lecture halls to voting booths."[10]

In the winter of 2020, the physicist Chanda Prescod-Weinstein published an article in *Signs* titled "Making Black Women Scientists under White Empiricism." Echoing the writings of Keller and Hammonds, she asked who is allowed to be an observer in physics and who is denied that possibility. She argued that scientists and academia continue to question the ability of Black students to be objective observers because their "standpoint on racism is different from that of white students and scientists who don't have to experience its consequences." Prescod-Weinstein used critical race theory, feminist standpoint theory, and intersectionality to reveal the racist and sexist gatekeeping in physics and to show that white supremacy in physics keeps Black physicists permanently in the role of other. In her view, making aggressive behavior a requirement for academic success in physics is especially harmful to Black women, who "are demonized for engaging in behaviors that even hint at aggression." Her article cites "The "Double Bind" report as well as writings by Hammonds, Keller, Subramaniam, Sandra Harding, Barbara Smith of the Combahee River Collective, and many others.[11]

In 2021, Prescod-Weinstein published *The Disordered Cosmos*, which won the *Los Angeles Times* award for science. In it, she urged readers to recognize that science is rife with racism, misogyny, and other forms of oppression. Yet, she declares, it is the fundamental right of all children to enjoy science, explore the nature of the world, and know the night sky.[12]

Feminist science scholarship has developed remarkably during the past fifty years. Because it does not operate as a single field with fixed disciplinary boundaries, its growth has not been, and will never be, linear. Often the genealogies of academics and scientists are described according to patriarchal measures of success: "how many graduate students did you supervise?" is an echo of "how many babies did you make?" But the impact of feminist science studies and feminist critiques of science has been more horizontal, more lateral, more viral. The ideas, methodologies, tools, and new knowledge will continue to spread far beyond their original sources.[13]

One example is the formation of Free Radicals, a collective founded in 2015 and dedicated, like Science for the People, to creating a more socially just, equitable, and accountable science. Inspired by Harding's work on strong objectivity, one of its co-founders, Sophie Wang, subsequently created *Science under the Scope*, a comic zine that explores questions of objectivity and science.[14]

In 2021, Emily Martin, the author of the classic essay "The Egg and the Sperm," spoke of how encouraged she's been by the number of undergraduate pre-meds and pre-scientists who are taking classes in feminist approaches. She said, "It could be that this kind of change takes a long time—getting an advanced degree, building a career secure enough to risk the kind of thing Ruth [Hubbard] did—all that could actually take thirty years." But as society shifts, Martin holds out hope for more rapid change.[15]

Feminists science studies still has no unified association or annual conference to bring together interested scientists and feminist scholars. However, in 2015, a group of feminists, including Banu Subramaniam, founded the international, peer-reviewed, open-source online journal *Catalyst: Feminism, Theory, Technoscience*.[16] Published twice a year, it operates on an independent platform outside of any university or press. Then, in 2016, Subramaniam and Rebecca Herzig began editing a feminist technoscience book series at the University of Washington Press. Both the series and *Catalyst* were expressly conceived as spaces to support feminist science and technology studies.

In 2020, *Catalyst* won the STS Infrastructure Award from the Society for the Social Studies of Science; and in April 2021, when Covid was still making in-person conferences unsafe, the journal celebrated its fifth anniversary with an online gathering. In the opening discussion, Subramaniam asked Donna Haraway, "Has feminist science studies been a successful project?" Haraway replied that the field had foregrounded the fact that women were scientists, that the conditions of women who did science were intolerable, and that the conditions of women who were seen as other, either racially or in terms of class or sexuality, were even more difficult. She was pleased that feminist science studies operated across disciplines via multifaceted strategies, an approach that had allowed young scholars in postcolonial and decolonial studies to draw from its findings.[17] Yet Haraway acknowledged that few scholars of feminist science studies had any deep engagement with practicing scientists. Nor were many feminist scientists taking the time or using the language that was needed to bridge the divides between more mainstream scientists. A huge amount of damage had been done by the Sokal hoax, when "we were accused of anti-science ideology." Intersectionality was another area of failure: "People like me have been intentionally anti-racist, but in practice most of my networks remain white."[18]

In her remarks, Subramaniam argued that feminist science studies was having a powerful impact, yet she also saw that, again and again, feminist contributions were being erased from academic and scientific discourse. The work of feminist scientists was being appropriated but not acknowledged. Other speakers at the online gathering, a diverse group of women from Asia, Latin America, the United States, and Europe, pointed out that women scholars, trans scholars, queer scholars, scholars of color, and indigenous scholars who do not fit the expected scientific description often quickly disappear from citation networks and that such citations are critical for ensuring that the field and its practitioners do not vanish.

At the same event feminist scholar Moya Bailey expressed excitement about the future of feminist science studies. Drawing inspiration from the Combahee River Collective, she had founded the Black Feminist Health Science Studies Collective, whose first symposium was held in March 2021. There, Evelynn Hammonds spoke about the importance of using the lens of feminist science studies to look at Black women's health, and Bailey spoke about the future for young women in this field. There is much work to be done, and these young scientists and science studies scholars will bring a vital feminist and antiracist awareness to their work. The seeds planted long ago are continuing to bear fruit.[19]

APPENDIX I

WHERE ARE THEY NOW?

Note: These brief bios are not complete. They are intended to offer selected information not covered fully in the book.

ROGER ARLINER YOUNG (1899–1964) taught at the North Carolina College for Negroes in Durham after completing her PhD in zoology at the University of Pennsylvania in 1940. Young joined the NAACP, and throughout the 1940s, she worked as a civil rights activist in voter registration and labor organizing, taking a job as chair of the biology department at Shaw University in Raleigh. In the 1950s, she taught at Shaw, Jackson State College, and several other colleges in Texas, Louisiana, and Mississippi. Young struggled with finances and her health, and died in November 1964 at the age of 65. In 2015, the Ocean Conservancy named the RAY Diversity Fellowship in Young's name to encourage underrepresented students in the ocean sciences.

BARBARA MCCLINTOCK (1902–1992) received many accolades for her decades of work in maize genetics after receiving the Nobel Prize in 1983. She continued as a researcher at Cold Spring Harbor until her death in 1992 at the age of 90. In 2001, Nathaniel Comfort published a second biography of McClintock titled *The Tangled Field: Barbara McClintock's Search for the Patterns of Genetic Control*, and in 2005 the US Postal Services issued a stamp with her image. In 2022, Cornell University opened McClintock Hall in her honor.

DOROTHY BURNHAM (born 1915) is 109 years old and lives in Brooklyn. For many years, she taught biology, bioethics, and health sciences in the adult education program at Empire State College. While she has not been widely recognized for her science contributions, she has received many awards for her role in public education, the civil rights movements, and other lifetime achievements.

RUTH HUBBARD (1924–2016) helped to found the Council for Responsible Genetics and served actively on its board. She was concerned that scientists wanted to link every human trait, disease, and behavior to a genetic cause, which would undermine the complexity of environmental factors. "The myth of the all-powerful gene" she wrote, "is based on flawed science that discounts the environment in which we and our genes exist."[1] She was a social activist and spent years writing and speaking about genetics, gender, and science, even after retiring from Harvard. Hubbard died at the age of ninety-two.

RITA ARDITTI (1934–2009) spent much of her teaching career at Union Graduate School, where her doctoral students worked on an array of inter-disciplinary topics that at times included the sciences. She also focused on women's health, cancer prevention, and human rights. In 1999, she published *Searching for Life: The Grandmothers of Plaza de Mayo and the Disappeared Children of Argentina.* One of her great inspirations was the biologist and environmentalist Rachel Carson. Arditti died at the age of seventy-five, after living with metastatic breast cancer for thirty-five years.

AUDRE LORDE (1934–1992), Black, feminist, lesbian, poet, mother, continued to speak, write, and publish until her death. Her many books include *Coal* (1976), *The Black Unicorn* (1978), *The Cancer Journals* (1980), *Zami: A New Spelling of My Name* (1983), *Sister Outsider* (1984), and *The Collected Poems of Audre Lorde* (1997). Lorde was the New York State Poet Laureate in 1991–1992. Three documentary films have been made about Lorde's life, including *A Litany for Survival* (2002), *The Edge of Each Other's Battles* (2002), and *Audre Lorde: The Berlin Years* (2012). Lorde's influence and legacy is broad across Black feminism and beyond, and there are many awards and projects in her honor.

SANDRA HARDING (born 1935), feminist and philosopher of science, was based at the Department of Philosophy and the Women's Studies Program at the University of Delaware from 1976–1996. She was the Director of the UCLA Center for the Study of Women from 1996–2000 and co-edited *Signs: Journal of Women in Culture and Society* from 2000–2005. She is the author and editor of many books including *Whose Science? Whose*

Knowledge?: Thinking from Women's Lives (1991) and *Is Science Multicultural? Postcolonialisms, Feminisms, and Epistemologies* (1998). Harding has been a consultant to the United Nations Commission on Science and Technology for Development and the U.N. Development Fund for Women. She is now retired and is a Distinguished Professor Emeritus of Education and Gender Studies at UCLA and a Distinguished Affiliate Professor of Philosophy at Michigan State University. She lives in Northampton, Massachusetts.

EVELYN FOX KELLER (1936–2023) retired from MIT and then published *The Mirage of a Space between Nature and Nurture* (2010). She received many honors for her work, including a MacArthur Fellowship (1992) on language, gender, and science. In 2018, she won the Israeli Dan David Prize, which she donated to organizations defending human rights in Palestine. In 2023, she published *Making Sense of My Life in Science: A Memoir.* She died at the age of eighty-five.

SHIRLEY MALCOM (born 1946) co-authored *The Double Bind: The Price of Being a Minority Woman in Science* in 1976, became the head of the AAAS Office of Opportunities in Science in 1979, and has continued to hold various leadership roles at AAAS ever since. In 1993, she was appointed to the National Science Board, and in 1995 she became a fellow of the American Academy of Arts and Sciences. Malcom has served as co-chair of the Gender Advisory Board of the United Nations Commission on Science and Technology for Development and has chaired many national committees on scientific education and literacy. She holds more than fifteen honorary degrees, and lives in Atlanta, Georgia.

NANCY HOPKINS (born 1943) retired from MIT and now divides her time between Massachusetts and New York City. In 2007, she married Dinny Adams, a New York attorney. She continues her work with MIT to ensure equity for women in the biotech industry via the Boston Biotech Working Group. Hopkins was among the MIT women science faculty members who were featured in the documentary *Picture a Scientist* (2020) and in Kate Zernike's book *The Exceptions* (2023).

MARGARET ROSSITER (born 1944), historian of science, published her landmark *Women Scientists in America: Struggles and Strategies to 1940* in 1982. In

1986, Rossiter had a one-year appointment at Cornell University which was extended and divided across three departments: women's studies, agriculture, and history, and in 1989 Rossiter was named a MacArthur Fellow. Only after a tenure offer from the University of Georgia was Rossiter's position at Cornell secured. She published Volume II, *Women Scientists in America: Before Affirmative Action, 1940–1972* in 1995. From 1994–2003, she was the editor of *Isis*, the official journal of the History of Science Society, and she published Volume III, *Women Scientists in America: Forging a New World Since 1972* in 2012.

ANNE FAUSTO-STERLING (born 1944) retired from Brown University in 2014 and holds the title of Nancy Duke Lewis Professor Emerita of Biology and Gender Studies. She is a fellow of the American Association for the Advancement of Science. One of her goals is to broaden the understanding of the inseparability of nature and nurture, and she continues to work to integrate feminist scholarship into science curriculums. Well known for her ability to explain scientific concepts to a lay audience, she writes frequently for the *Boston Review*, the *Huffington Post*, and *Psychology Today*. She published *Sex/Gender: Biology in a Social World* in 2012, a second edition of *Sexing the Body* in 2020, and is working on a new book.

SHIRLEY ANN JACKSON (born 1946), physicist, joined Bell Laboratories in 1976, where she worked for fifteen years. She served on the Rutgers University faculty from 1991–1995, the year she was appointed as Chair of the Nuclear Regulatory Commission. In 1999, Jackson became the 18th president of Rensselaer Polytechnic Institute (RPI) in New York, a position she held until 2022. She has received many awards and honors, including the National Medal of Science in 2014.

BARBARA SMITH (born 1946) was the co-founder of Kitchen Table: Women of Color Press and remains a widely known feminist, social activist, writer, and public speaker. Smith edited *Home Girls: A Black Feminist Anthology* (1983) and authored *The Truth that Never Hurts: Writing on Race, Gender and Freedom* (1998). She lives in Albany, New York and served on the Albany City Council between 2005 and 2013.

BEVERLY SMITH (born 1946) has retired from her work in women's public health and lives in Massachusetts. In addition to her Masters in Public

Health from Yale, she has a master's from the Harvard Graduate School of Education. She stays active as a Black feminist health advocate and her essays and articles on feminism, racism, and women's health have been published widely. She has also been involved in Unitarian Universalism for more than thirty years.

DEMITA FRAZIER (born 1950), co-founder of the Combahee River Collective, graduated from Northeastern University School of Law in 1986. Frazier has continued as a Black feminist, social justice activist, writer, and teacher on issues including reproductive rights, domestic violence, and environmental justice in poor and working-class communities of color. She has taught throughout the Boston area and New England.

EVELYNN HAMMONDS (born 1953) is the Barbara Gutmann Rosenkrantz Professor of the History of Science at Harvard, and she chaired the history of science department from 2017 to 2022. She is also a professor of African and African American studies and a professor of social and behavioral sciences at the Harvard's Chan School of Public Health. Hammonds served as dean of Harvard College from 2008 to 2013. She was elected to the National Academy of Medicine in 2018 and to the American Academy of Arts and Sciences in 2021. She is president of the History of Science Society. And in 2022, she became the inaugural Audre Lorde Professor of Queer Studies at Spelman College in Atlanta.

BANU SUBRAMANIAM (born 1966) is the Luella LuMer Professor and chair of the Department of Women's and Gender Studies at Wellesley College in Wellesley, Massachusetts. She is the author of three books, including the award-winning *Ghost Stories for Darwin: The Science of Variation and the Politics of Diversity* (2014), *Holy Science: The Biopolitics of Hindu Nationalism* (2019), and *Botany of Empire: Plant Worlds and the Scientific Legacy of Colonialism* (2024).

APPENDIX 2

DECLARATION

Equality for Women in Science

This statement and its proposed resolutions were presented by a group of women scientists, including Rita Arditti, at the 1969 meeting of the American Association for the Advancement of Science (AAAS). The association did not adopt the resolutions, and the editors of *Science* declined to publish the statement. The declaration did, however, appear in the first issue of *Science for the People* (August 1970).[1]

The stated goals of the American Association for the Advancement of Science (AAAS) are:

To further the work of scientists,

To facilitate cooperation among them,

To improve the effectiveness of science in the promotion of human welfare, and

To increase public understanding and appreciation of the importance and promise of the methods of science in human progress.

None of these objectives can be realized while women in science are relegated to second class status. Female scientists do not escape the oppression faced by all women in our society. They are oppressed economically and culturally—trained for inferior roles and exploited as sex objects and consumers.

Such sexual discrimination is no accident. It serves, in a variety of ways, the interests of those who dominate the economy of this country. It provides them with a source of ideologically justified cheap labor, and as a consequence drives all wages down. It establishes "wives" as unpaid household workers and child raisers, as well as a body of willing consumers. At the same time, the limitations on the creative development of women deprive society of the full contributions of over one half [of] its members.

It is important to note that sexual oppression is both pervasive and institutionalized; within the scientific community it takes many forms. Educational tracking by sex from elementary school on channels women into subordinate roles and stereotypes. While men are trained to develop "logical" patterns of thought, women are encouraged to be "intuitive." Math and science are seen as male prerogatives. Vocational counselling in high schools and colleges pressures women into family roles, clerical work and, if professions are considered, into the service fields: teaching, social work, nursing, etc. Those few women who manage to transcend such socialization and choose scientific careers, encounter a vicious circle of exploitation. Quotas are placed on graduate school admissions and justified by the self-fulfilling prophecies that most women will be unable to finish because they will marry, have children, and lack the emotional stability and drive to meet the arduous initiation rites of the profession. The still fewer who complete their training continue to find themselves between family and profession, while men never have to make that choice.

As scientists, they are limited by being placed in subordinate positions, rarely being given their own labs or first authorship on papers, and, the most glaring inequity, being paid less than their male colleagues for equal work. They are automatically and illegally barred from certain jobs, particularly in industry, and cut off from tenured and supervisory positions.

Moreover, the psychological harassment is constant and debasing. Casual remarks continually define the female scientist simply in relation to her sex, from compliments on her looks to "you think like a man." She is placed in the schizophrenic position of being treated as either a dehumanized worker or a feminine toy.

Universities hold a strategic position with regard to all manifestations of this problem, since they help create and transmit the ideology of male supremacy.

Moreover, the practices of sexual discrimination which permeate all institutions where AAAS members work and study are contradictory to the declared goals of the AAAS. Clearly we cannot "further the work of scientists" while denigrating in so many ways the contributions and potential of women in the profession. Sexual discrimination make "cooperation among scientists" an ironic platitude. The "effectiveness of science in the promotion of human welfare" is hardly furthered by denying half of humanity the opportunity to pursue scientific careers, or by wasting this tremendous reservoir of talent.

We therefore propose the following resolutions be adopted at the general Council meeting of the AAAS, and be fought for by AAAS members where they work.

1. That universities and other institutions where AAAS members work be immediately required to comply with the law of the land and pay equal wages for equal work to women and men.
2. That graduate school departments and medical schools admit ½ women and ½ men, regardless of the proportion of applicants, and that they take whatever steps are necessary to recruit sufficient women to comply with this demand.
3. That vocational counselling in high schools and colleges be totally reoriented so as not to channel women into low-status, low potential occupations.
4. That universities and other institutions give priority to the hiring and promotion of women, increasing the proportion of women to 50% at all levels.
5. That birth control and abortion counselling be provided by university and company health services to all women.
6. That the curriculum of courses in psychology, sociology, anthropology, etc., be thoroughly revamped by women, to end the perpetuation and creation of male supremacist myths.

    Further that sex inequality be added as a topic to all courses and texts which cover social inequalities, and that new courses be created by women about their history and oppression.
7. That universities and government sponsor programs to investigate and change the subordinate status of women in our society.
8. That it be recognized that the actual practices of hiring, promotion and tenure discriminate against women, and that institutions have not accepted their responsibility for such inequalities.

As a first step in the right direction institutions should provide:

a. parenthood leave and family sick leave for all employees, female and male.
b. half-time appointments for mothers and fathers who want them must be considered. (Since child-rearing is a social responsibility, it is preferable for both parents' work to be slowed down than for the mother's to be stopped entirely.)
c. free child care centers should be provided for all children. These day care centers should be open to the communities where the institutions are located, controlled by the parents, staffed equally by male and female teachers, open 24 hours a day, 7 days a week for infants to school age children and after school for older children.

While we realize that the ultimate liberation of both women and men in our society will only come with a total social and economic revolution, we feel that it is important for us to make steps now toward destroying false notions of superiority which do not serve science, scientists, or humanity.[2]

ACKNOWLEDGMENTS

I started working on *Our Science, Ourselves* in 2019, which meant that all of the women in this book kept me company throughout the isolating years of Covid. The idea of writing a book about people and events in Boston while living in Johannesburg, South Africa, seemed slightly crazy but doable. (Come to think of it, people do the reverse all the time.) Still, when I started, I had no idea that I would have to continue my research long-distance for so long. I am thankful to the many people who encouraged me to keep going.

Thank you to the Wits Institute for Social and Economic Research (WiSER) for giving me positive feedback on my initial proposal and offering me a base from which to conduct research. The opportunity to present my work to WiSER internal seminars in August 2019 and October 2021 helped guide my progress. Thank you to Sarah Nuttall, Hlonipha Mokoena, Keith Breckenridge, Pamila Gupta, Richard Rottenburg, Jonathan Klaaren, Najibha Deshmukh, Adila Deshmukh, Terry Kurgan, and Sizwe-Mpofu Walsh. Thank you to Souad Zeineddine and Jia-Hui Lee for pointing me in fruitful directions.

In 2019, Jane Kamensky and Harvard University's Schlesinger Library on the History of Women in America selected me for the Joan Challinor Award, a research grant. Thank you to Jennifer Fauxsmith and everyone at the Schlesinger for always being supportive and helpful. The archive holds countless papers and audio resources that benefited my research.

Thank you to the Consortium for History of Science, Technology, and Medicine (CHSTM) for selecting me as a 2019–20 research fellow. Not only did the grant support my research, but the CHSTM was an online lifeline to the history of science community in the United States. Thank you to Babak

Ashrafi and Lawrence Kessler and to everyone who attended my online presentation in November 2020.

I want to thank everyone who took part in interviews, discussions, and email exchanges: Evelynn Hammonds, Banu Subramaniam, Evelyn Fox Keller, Anne Fausto-Sterling, Nancy Hopkins, Beverly Smith, Sandra Harding, Dolita Cathcart, Nancy Krieger, Erika Milam, Jonathan Beckwith, Barbara Beckwith, Elijah Wald, Deborah Wald, Federico Muchnik, Estelle Disch, Judith Masters, Mila Pollock, Judy Norsigian, Judith Walzer, Emily Martin, Sandra Korn, Peggy McIntosh, Ursula Goodenough, Abha Sur, Betsey Useem, Bonnie Spanier, Marian Lowe, Linda Burnham, Dorothy Burnham, Nayanika Ghosh, Nadja Cech, Karen Martin, Femke Brandt, and Simonne Horwitz. I valued my exchanges with each of you, as you helped me to understand scientific concepts, piece together events in history, and just keep going.

Thank you to Wendy Call, who led an advanced creative nonfiction workshop for *Writers.com*. In 2021, I shared excerpts of my manuscript with classmates Heather White, Nadya Cech, Vicki Clayton, Jeremy Simer, Tara Bilboy, Jessica Kulynich, and Valerie Ashton, all of whom offered helpful feedback. Special thanks to Wendy Call for reading my entire draft manuscript. Her editorial suggestions and feedback helped me keep working on the book at a time when it felt like a steep mountain to climb.

Ann Robinson assisted me at Harvard's Widener Library, and Robin McElheny welcomed me to the Harvard University Archives and took interest in my project. Thank you to Amanda Knox, who helped me prepare for my visit to the John Hay Library at Brown University. Mila Pollock helped me with long-distance searches at the Cold Spring Harbor Archives. Holly A. Smith, the archivist at the Women's Research and Resource Center at Spelman College, gave me enthusiastic long-distance help with a search of Audre Lorde's papers. I am grateful to the online archives at the Massachusetts Institute of Technology and Science for the People.

Thank you to everyone who helped me with photographs: Janet Van Ham, Kathy Seltzer, Starr Ockenga, Laura Wulf, Matthew Tontonoz, Estelle Disch, Federico Muchnik, Elijah Wald, Deborah Wald, Linda Burnham, Anne Fausto-Sterling, Jennifer Fauxsmith at Schlesinger Library, Christine Daniloff at MIT News, Ariel Weinberg at the MIT Museum, and Calvin Wu at Science for the People.

I'm grateful to Nayanika Ghosh, Beverly Smith, Pamila Gupta, and Alice Brown for reading selected chapters. And a big thank you to Nadja Cech for reading sections of the manuscript for scientific accuracy and helping me to portray the science in more compelling ways. I greatly appreciate Kate Zernike's willingness to share the results of her research regarding Ruth Hubbard's and George Wald's joint Paul Karrer Prize.

Many thanks to Banu Subramaniam, Evelynn Hammonds, Nancy Hopkins, Ken Manning, Dana-Ain David, Makhosazana (Khosi) Xaba, and Roger Jardine for reading the entire manuscript and giving me feedback. I'll always remember that indescribable moment with Khosi when the title came to us. And thanks to everyone who helped me sculpt the subtitle.

Alison Lowry has listened to me talk about this book from the start. She read early draft chapters, discussed structural changes, and let me know when the manuscript was starting to hold together. It was Alison's edit that helped me feel that I could send it off to the publisher.

I am enormously grateful to Banu Subramaniam, not only for her early interest and support but also for suggesting that I send my book proposal to the University of Massachusetts Press. Thank you to Sigrid Schmalzer, a historian of science and the editor of the press's series Activist Studies of Science and Technology, for her support in revising important sections of my manuscript. Thank you to Matt Becker, the press's editor-in-chief, for his help with the introduction and for shepherding my manuscript through peer review. Thank you to the two anonymous peer reviewers for thoughtful feedback that has improved the book in numerous ways. I was thrilled to find out that accomplished poet and author Dawn Potter would copyedit the manuscript. *Our Science, Ourselves* has benefited enormously from her care and attention. Thank you to Sally Nichols and the production team for their wonderful support in bringing the book to press. Thank you to Sandra Sadow for the index, and thank you to Chelsey Harris for helping to spread the word.

I am thankful to my family and friends for supporting me with this project during a very difficult time—one of isolation and loss but also togetherness and love. Big thank yous to Jewel, Bob, and Sarah Kuljian, who supported me from the start, and to Anne Jardine, more widely known as Grandma Anne. I am grateful for the support of Fab Four; Lockdown Luvvies; Lamejun; Cousins, Aunts, and Uncles; Book Club; Familia; and the Jardine Tribe. A special thank you to Nadia and Mila and Roger Jardine, who are always there for me.

NOTES

INTRODUCTION

1 According to numerous obituaries, Ruth Hubbard received tenure in 1974. However, both Wikipedia and the *Harvard Crimson* list the date as 1973. My guess is that discussions began in 1973 and were confirmed in early 1974, so that is the date I will use in this book. See Bryan Marquard, "Ruth Hubbard, 92, First Woman Tenured in Biology at Harvard," *Boston Globe*, September 4, 2016; Jonathan Beckwith, Evelynn Hammonds, Nancy Krieger, and John Dowling, "Ruth Hubbard Wald, 92," *Harvard Gazette*, March 8, 2017; Sara Corbett, "Ruth Hubbard Challenged the Male Model of Science," *New York Times Magazine*, December 21, 2016; Harvard University, Department of Molecular and Cellular Biology, "Ruth Hubbard Wald, Harvard Biologist, Dies at 92," October 24, 2016, https://www.mcb.harvard.edu; and Stephen M. Gillinov, "Ruth Hubbard, Harvard Biology Professor and Political Activist, Dies at 92," *Harvard Crimson*, September 8, 2016.

2 It is common to write about scientists using last names only. Thus, in an effort to make sure that all scientists in this book, both women and men, are treated equally, I will use last names predominantly. However, I will selectively use first names when I write about scientists in the context of their childhood and youth or in very personal circumstances.

3 Ruth Hubbard, "Reflections on my Life as a Scientist," *Radical Teacher* 30 (1981): 6.

4 Ruth Hubbard, "Have Only Men Evolved?," in *Women Look at Biology Looking at Women: A Collection of Feminist Critiques*, ed. Ruth Hubbard, Mary Sue Henifin, and Barbara Fried (Cambridge, MA: Schenkman, 1979), 21.

5 Christa Kuljian, "Periodic Weakness: The Medical View of Menstruation, 1870–1920" (bachelor's thesis, Harvard University, 1984).

6 James Damore, "Google's Ideological Echo Chamber," https://www.documentcloud .org; Daisuke Wakabayashi, "Google Fires Engineer Who Wrote Memo Questioning Women in Tech," *New York Times*, August 7, 2017; Maryam Zaringhalam and Jess Wade, "A New Front in Fighting the Bias against Women in Science," *Scientific American*, September 17, 2018, http://scientificamerican.com.

7 More than 4,000 scientists from around the globe signed a statement pointing out the scientific and moral problems with Strumia's remarks. It read, in part: "Physics and science are part of the shared inheritance of all people, as much as art, music and

literature, and we should strive to ensure that everyone has a fair opportunity to become a scientist" (Particles for Justice, "Statement on Sexism," 2018, https://www.particlesfor justice.org).

8 Lawrence Summers, "Full Transcript: President Summers's Remarks at the National Bureau of Economic Research, January 14, 2005," *Harvard Crimson*, February 18, 2005.

9 In an oral history interview produced and directed by Joan Rachlin and filmed by John Rich at the home of Margaret Burnham in Jamaica Plain, Massachusetts, in October 2014, Dorothy Burnham said, "I finished a master's degree and started working in the laboratories in St. John's Hospital in Brooklyn and at Staten Island Hospital and at Sloan Kettering." Margaret Burnham and Linda Burnham, Dorothy's daughters, kindly allowed me to watch the entire four-and-a-half-hour film, a small portion of which is available online (https://vimeo.com/118977130). Unfortunately, despite years of searching, I have not been able to confirm the date and institution of Burnham's master's degree.

10 Rachlin and Rich, film about Burnham; Dorothy Burnham, "Biology and Gender," in *Genes and Gender*, ed. Ethel Tobach and Betty Rosoff (New York: Gordian, 1978), 51–59; Christa Kuljian, interview with Dorothy Burnham and Linda Burnham, August 24, 2021.

11 Patricia Hill Collins, "Moving beyond Gender: Intersectionality and Scientific Knowledge," in *Revisioning Gender*, ed. Myra Marx Ferree, Judith Lorber, and Beth B. Hess (Walnut Creek, CA: Altamira, 2000); and Kimberlé Crenshaw, "Demarginalizing the Intersection of Race and Sex: A Black Feminist Critique of Antidiscrimination Doctrine, Feminist Theory, and Antiracist Politics," *University of Chicago Legal Forum* (1989): 139–67.

CHAPTER I

1 Londa Schiebinger, "Maria Winkelmann at the Berlin Academy: A Turning Point for Women in Science," *Isis* 78 (June 1987): 174–200; Londa Schiebinger, *The Mind Has No Sex? Women in the Origins of Modern Science* (Cambridge, MA: Harvard University Press, 1989); David Noble, *A World without Women: The Christian Clerical Culture of Western Science* (New York: Knopf, 1993).

2 Margaret Rossiter, *Women Scientists in America*, vol. 1, *Struggles and Strategies to 1940* (Baltimore: Johns Hopkins University Press, 1982), 9.

3 Rossiter, *Women Scientists in America*, 1:xv.

4 Charles Darwin, *The Descent of Man* (London: Penguin Classics, 2004), 629–31.

5 Caroline Kennard, letters to Charles Darwin, December 26, 1881, January 28, 1882, and Charles Darwin, letter to Caroline Kennard, January 9, 1882, all in Darwin Correspondence Project, Cambridge University, https://www.darwinproject.ac.uk.

6 Rossiter, *Women Scientists in America*, 1:xv.

7 Rossiter, *Women Scientists in America*, 1:30–31.

8 Rossiter, *Women Scientists in America*, 1:14.

9 Rossiter, *Women Scientists in America*, vol. 1, chaps. 6 and 7.

10 R. A. Young, "On the Excretory Apparatus in *Paramecium*," *Science* 60 (1924): 244; Sara P. Díaz, "'A Racial Trust': Individualist, Eugenicist, and Capitalist Respectability in the Life of Roger Arliner Young," *Souls* 18 , no. 2–4 (2016): 235–62; Sara P. Díaz, "Gender, Race, and Science: A Feminista Analysis of Women of Color in Science" (PhD diss., University of Washington, 2012); Ken Manning, *Black Apollo of Science* (Oxford: Oxford University Press) 1983).

11 Rossiter, *Women Scientists in America*, 1:175.

12 Naomi Oreskes, "Objectivity or Heroism? On the Invisibility of Women in Science," *Osiris* 11 (1996): 87–113.

13 Margaret Rossiter, *Women Scientists in America*, vol. 2, *Before Affirmative Action, 1940–1972* (Baltimore: Johns Hopkins University Press, 1989), 188–90.

14 *American Men of Science* was finally changed to *American Men and Women of Science* in 1971.

CHAPTER 2

1 Margaret Rossiter, *Women Scientists in America*, vol. 2, *Before Affirmative Action, 1940–1972* (Baltimore: Johns Hopkins University Press, 1995), xv–xviii.

2 Judith Walzer, interview with Ruth Hubbard (1981), 42, Judith Walzer Papers, Schlesinger Library, Harvard University; Laurel Ulrich, ed., *Yards and Gates: Gender in Harvard and Radcliffe History* (New York: Palgrave Macmillan, 2004), 342.

3 Rossiter, *Women Scientists in America*, 2:xxx; Ulrich, *Yards and Gates*, 344. On Hubbard's experiences with poor teaching at Radcliffe, see ibid., 341–44.

4 Marguerite Holloway, "Profile: Ruth Hubbard—Turning the Inside Out," *Scientific American* 272, no. 6 (1995): 49; Walzer, interview, 50–51.

5 Walzer, interview, 6, 15–20.

6 Walzer, interview, 22–24.

7 Walzer, interview, 26–28.

8 Ruth Hubbard and Margaret Randall, *The Shape of Red: Insider/Outsider Reflections* (San Francisco: Cleis, 1988), 64,

9 Holloway, "Profile: Ruth Hubbard," 49; Madeline Drexler, "Ruth Hubbard, an Outsider Inside," *Boston Globe*, May 16, 1990.

10 Walzer, interview, 52, 36; Ruth Hubbard, personal journal, April 14, 1976. I'm grateful to Elijah Wald for sharing the journal with me.

11 Walzer, interview, 46, 61; Bonnie Spanier, interview with Ruth Hubbard (1986), 1–6, Bonnie Spanier Papers, Schlesinger Library, Harvard University.

12 Donald R. Griffin, Ruth Hubbard, and George Wald, "The Sensitivity of the Human Eye to Infra-Red Radiation," *Journal of the Optical Society of America* 37 (1947): 546–54; Walzer, interview, 60–61; Spanier, interview with Hubbard, 1–7, 1–12; "Ruth Hubbard," *How to Think about Science* (Ottawa: Canadian Broadcasting Corporation, 2007).

13 Walzer, interview, 66; Spanier, interview with Hubbard, 1–8, 1–9.

14 Ulrich, *Yards and Gates*, 345.

15 Spanier, interview with Hubbard, 1–11.

16 Walzer, interview, 73. On Frank Hubbard, see Michael Steinberg, "Habit of Perfection: A Tribute—Obituary to Frank Hubbard," *Hubbard Harpsichords*, February 1976, http://www.hubharp.com.

17 Walzer, interview, 76.

18 Walzer, interview, 78–79.

19 Ruth Hubbard, "The Logos of Life," in *Reinventing Biology: Respect for Life and the Creation of Knowledge*, ed. Lynda Birke and Ruth Hubbard (Bloomington: Indiana University Press, 1995), 95.

20 Hubbard, "The Logos of Life," 95.

21 Paul Brown, comment, in "In Celebration of Ruth Hubbard," symposium, Harvard University, June 2, 1990, box 13, folder 6, Ruth Hubbard Papers, Schlesinger Library, Harvard

University; Allen Kropf, comment, in ibid.; Beryl Lieff Benderly, "Ruth Hubbard and the Evolution of Biology," *Science*, October 5, 2016. The symposium took place on the occasion of Hubbard's appointment as professor emerita at Harvard.

22 Walzer, interview, 84, 136.

23 Evelyn Fox Keller, "The Anomaly of a Woman in Physics" (1977), in *Women, Science, and Technology: A Reader in Feminist Science Studies*, ed. Mary Wyer, Mary Barbercheck, Donna Geisman, Hatice Orun Ozturk, and Marta Wayne (New York: Routledge, 2001), 10–11. Keller tells this story in her essay without identifying names. However, her draft of *Making Sense of My Life in Science: A Memoir*, which she shared with me in 2019, reveals them to be Roy Glauber and Sheldon Glashow. Glauber won the Nobel for his work in quantum optics in 2005. Glashow won a Nobel for physics in 1979.

24 Keller, "The Anomaly," 10; Christina Agapakis, "Conversations with Evelyn Fox Keller," *Method Quarterly* (2014), http://www.methodquarterly.com.

25 Agapakis, "Conversations."

26 Agapakis, "Conversations."

27 Evelyn Fox Keller, *Making Sense of My Life in Science: A Memoir* (Amherst, MA: Modern Memoirs, 2023), 54–55.

28 Keller, *Making Sense of My Life in Science*, 43–45.

29 Keller, "The Anomaly," 13; Evelyn Fox Keller, "Women in Science: An Analysis of a Social Problem," *Harvard Magazine* (October 1974): 17.

30 Keller, "The Anomaly," 13.

31 Keller, "The Anomaly," 14.

32 Evelyn Fox Keller, *A Feeling for the Organism: The Life and Work of Barbara McClintock* (New York: Freeman, 1983).

33 Catherine Russo, video interview with Rita Arditti (1991), https://ritaarditti.com; Bonnie Spanier, interview with Rita Arditti (1978), box 1, folder 6, Bonnie Spanier Papers, Schlesinger Library, Harvard University.

34 Bonnie Spanier, interview with Arditti (1978).

35 Spanier, interview with Arditti (1978); Estelle Disch, personal communication, September 2, 2023.

36 Spanier, interview with Arditti (1978); Rita Arditti and Giuseppe Sermonti, "Modification by Manganous Chloride of the Frequency of Mutation Induced by Nitrogen Mustard," *Genetics* 47 (June 1962): 761–68; Rita Arditti and Anna Coppo, "Effect of Acridines and Temperature on a Strain of *Bacillus magaterium* Lysogenic for Phage x," *Virology* 25 (1965): 643–49.

37 Rita Arditti, "Feminism and Science," in *Science and Liberation*, ed. Rita Arditti, Pat Brennan, and Steve Cavrak (Boston: South End, 1980).

38 Arditti, "Feminism and Science."

39 Spanier, interview with Arditti, (1978).

40 Jon Beckwith, *Making Genes, Making Waves: A Social Activist in Science* (Boston: Harvard University Press, 2002), 71–71; Christa Kuljian, interview with Barbara Beckwith and Jon Beckwith, June 15, 2019.

41 Beckwith, *Making Genes, Making Waves*, 72.

42 Thank you to the chemist Nadja Cech for her personal communication, July 4, 2023.

43 Christa Kuljian, interview with Federico Muchnik, September 18, 2019.

44 Kuljian, interview with Muchnik; Federico Muchnik, personal communication, August 23, 2023.

45 Audio interview with Rita Arditti (c. 1979–80), interviewer unknown, collection of Federico Muchnik; Estelle Disch, "Rita Arditti," *FemBio*, http://fembio.org.

46 Rita Arditti, "My Ancestors Speak," in *The Tribe of Dina: A Jewish Women's Anthology*, ed. Melanie Kaye Kantrowitz and Irena Klepfisz (Boston: Beacon, 1989); audio interview of Arditti.

47 Walzer, interview, 91.

48 Walzer, interview, 96–97.

49 Walzer, interview, 92.

50 After Zernike wrote to the University of Zurich about Hubbard's missing name, Karl Gademann, the administrator of the award, found Wald's March 8, 1967, letter and updated the records. See Kate Zernike, *The Exceptions: Nancy Hopkins, MIT, and the Fight for Women in Science* (New York: Scribner, 2023), 68, n. 376; and Kate Zernike, personal communication, April 18, 2023.

51 Kropf, comment; Spanier, interview with Hubbard, 1–19.

52 Walzer, interview, 93–95; Christa Kuljian, interview with Deborah Wald, June 25, 2020; Carl Cobb, "Profile in the News: Master Teacher," *Boston Globe*, October 19, 1967.

53 Walzer, interview, 95. While Hubbard did not reveal Theorell's name in her interview with Walzer, she did write about the incident in her personal journal, dated summer 1970: "Theorell's comment on Nobel prize looms inordinately large. . . . Except for the Theorell remark, I have no associated feeling of trauma." Thank you to Elijah Wald for sharing the journal.

54 Walzer, interview, 95.

55 See Anne Sayre, *Rosalind Franklin and DNA* (New York: Norton, 1975); and Hubbard's untitled review of the book in *Signs* 2 (August 1976): 229–37.

56 Walzer, interview, 93.

57 Walzer, interview, 91; Margaret Rossiter, "The Matthew/Matilda Effect in Science," *Social Studies of Science* 23 (May 1993): 330. Rossiter quotes the Walzer interview and cites Patricia Farnes, "Women in Medical Science," in *Women of Science: Righting the Record*, ed. Gabriele Kass-Simon and Patricia Farnes (Bloomington: Indiana University Press, 1993), 289, n. 11. Walzer's interview appears to be the only time when Hubbard shared her thoughts on the record.

58 Rossiter, "The Matthew/Matilda Effect," 330.

CHAPTER 3

1 Janet Christensen, "12 Intruders Protest Women's 'Inferior Role,'" *Boston Herald-Traveler*, December 30, 1969; Robert Sales, "A Women's Place in Science," *Boston Globe*, December 30, 1969. See appendix 2 for the full text of the eight resolutions and the overall declaration.

2 An article by Madeline Drexler, on the occasion of Hubbard's retirement, correctly mentions the 1969 protest and Hubbard's "dawning of consciousness" but wrongly refers to the "American Academy of Arts and Sciences meeting in Boston" rather than the American Association for the Advancement of Science meeting ("Ruth Hubbard, an Outsider Inside: A Harvard Biologist Challenges the Way Science Looks at the World," *Boston Globe*, May 16, 1990). See Bonnie Spanier, interview with Ruth Hubbard (1986), box 2, 1–19, Bonnie Spanier Papers, Schlesinger Library, Harvard University; and Judith Walzer, interview with Ruth Hubbard (1981), Judith Walzer Papers, Schlesinger Library, Harvard University.

3 Sales, "A Woman's Place in Science."

4 Christensen, "12 Intruders"; Sales, "A Women's Place in Science."

5 Drexler, "Ruth Hubbard, an Outsider Inside."

6 The 1969 conference was held in the same city where the AAAS had its first meetings in 1847 before being founded in Philadelphia the following year. The newly built War Memorial Auditorium and the Sheraton Boston Hotel near the Prudential Center in Back Bay housed the ninety-three conference sessions planned over five days. See Victor McElheny, "10,000 Scientists Gather," *Boston Globe*, December 25, 1969; and "Channel 2's 25 Hours of Science," *Boston Globe*, December 25, 1969.

7 Audio interview with Rita Arditti (1979–80), interviewer unknown; Catherine Russo, video interview with Rita Arditti (1991), https://ritaarditti.com.

8 Audio interview with Arditti.

9 For the full text of Wald's speech, see http://elijahwald.com.

10 Frank Rich, "Echoes of 1969: Recalling a Time of Trial, and Its Continuing Resonances," *Harvard Magazine* (March–April 2019), https://www.harvardmagazine.com.

11 On the early history of SSPA, SESPA and the formation of Science for the People, see "History of SESPA," *Science for the People* 2 (December 1970): 2–3, and Kelly Moore, *Disrupting Science* (Princeton, NJ: Princeton University Press, 2008), 146–57.

12 For a complete history, see Kathy Greeley and Sue Tafler, "Science for the People—A Ten Year Retrospective," *Science for the People* 11 (January–February 1979): 18–25.

13 Jon Beckwith, *Making Genes, Making Waves: A Social Activist in Science* (Cambridge, MA: Harvard University Press, 2002) 72.

14 Moore, *Disrupting Science*, 161–63.

15 Russo, video interview with Arditti.

16 Russo, video interview with Arditti.

17 Bonnie Spanier, interview with Rita Arditti (1978), box 1, folder 6, Bonnie Spanier Papers, Schlesinger Library, Harvard University.

18 "Rita Arditti, a Sephardic Human Rights Activist, died December 25, 2009 at Age 75," *eSefarad*, January 1, 2010, http://esefarad.com.

19 Rita Arditti, "My Ancestors Speak," *eSefarad*, April 1, 2010, https://esefarad.com; Estelle Disch, personal communication, September 2, 2023; "Rita Arditti, a Sephardic Human Rights Activist"; Jewish Women's Archive, http://jwa.org. Federico Muchnik, "Rita Arditti: Activist, Biologist, Teacher," video. In the 1960s, Arditti did not describe herself as a person of color or as a Third World woman, the term current at the time. However, according to her partner Estelle Disch, that later changed. After 9/11, several people, hearing her voice on the phone, asked , "Where are you from?" or told her to "Go back home."

20 Spanier, interview with Arditti (1978).

21 Moore, *Disrupting Science*, 163; Richard Knox, "Students to Examine 'Sorry State of Science'" *Boston Globe*, December 27, 1969; Jean Dietz, "Irate Women Speak Up for Day Care," *Boston Globe*, December 29, 1969.

22 Robert J. Sales, "Anti-Smokers and Soap Bars," *Boston Globe*, December 29, 1969; Victor K. McElheny, "AAAS Mulls Steps to Stop Disrupters," *Boston Globe*, December 28, 1969.

23 Elizabeth Fox-Wolfe created the iconic fist-and-flask logo for Science for the People. See Herb Fox, "Visualizing Radical Science: A Tribute to Elizabeth Fox-Wolfe, 1926–2019," *Science for the People* 22, no. 1 (2019), https://magazine.scienceforthepeople.org.

24 Moore, Disrupting Science, 164; Richard Knox, "AAAS Council Hauls Science out of White Tower," *Boston Globe*, December 31, 1969; Margaret Rossiter, *Women Scientists in*

America, vol. 2, *Before Affirmative Action, 1940–1972* (Baltimore: Johns Hopkins University Press, 1995), 372. According to Rossiter, Alice Rossi was more successful with her petition to the American Sociology Association in 1969–70. Instead of simply presenting resolutions at the annual conference, (1) she and others prepared a report on the status of women in the field, which they published before the meeting; (2) she presented the report to an open session early in the conference; and (3) she presented the resolutions to the governing council toward the end of the meeting. Her strategy was successful, and the council approved the resolutions.

25 Immediately after World War II, Mina Rees was the only woman scientist at the Office of Scientific Research and Development to hold a significant position. Even so, she was listed as a "technical aide" and an "executive assistant to the chief of the applied mathematics panel" (Rossiter, *Women Scientists in America*, 2:7). See Jean Dietz, "Scientists to Elect First Woman Leader," *Boston Globe*, December 30, 1969; and Wolfgang Saxon, "Mina S. Rees, Mathematician and CUNY Leader, Dies at 95," *New York Times*, October 28, 1997.

26 Knox, "AAAS Council Hauls Science out of White Tower."

27 Knox, "AAAS Council Hauls Science out of White Tower."

28 [Rita Arditti], "The Medical School," in *How Harvard Rules Women*, unnamed editors (Chicago: New University Conference, 1970), 36.

29 [Arditti], "The Medical School, 36.

30 Christa Kuljian, interview with Elizabeth Useem, September 23, 2019; Elizabeth Useem, personal communication, April 14, 2020. The pamphlet authors were affiliated with the New University Conference, a national organization working at universities to build socialism. A description of the organization appears in *How Harvard Rules Women*, 78.

31 Unnamed author, "Prologue," in *How Harvard Rules Women*, 3, quoted in "A Decision to 'Lance the Boil,'" *Boston Sunday Globe*, October 12, 1969.

32 [Arditti], "The Medical School," 37–40.

33 The Boston Marathon had excluded women since its founding in 1867. When Roberta Gibb applied to run officially in 1966, the race director's response was that women were "not physiologically capable of running 26 miles," so she ran unofficially. In 1967, Kathrine Switzer ran with an official number because she applied using only her initials. Once she was discovered on the road, one of the race managers physically attacked her. Not until 1972 were women officially allowed to run in the race. In 1975, Marilyn Bevans was the first African American woman to run it in under three hours. See Roberta Gibb, "A Run of One's Own," *Running Past*, n.d., http://runningpast.com; Kathrine Switzer, "WBZ Archives: Kathrine Switzer Makes History at the Boston Marathon," n.d., http://youtube.com; and Purcell Dugger, "'Just Run': Trailblazing Marathoner Marilyn Bevans Met the Challenges and Became a Champion," *Andscape*, November 2, 2018, http://andscape.com

34 Cold Spring Harbor Laboratory, list of participants, Symposium on Quantitative Biology, June 4–11, 1970, Cold Spring Harbor Laboratory Library and Archives; https://library.cshl.edu; Christa Kuljian, interview with Federico Muchnik, September 18, 2019. Ludmila Pollock, the executive director, of the Cold Spring Harbor Laboratory Library and Archives, kindly shared Rita Arditti's paper, "In Vitro Transcription of the *lac* Operon Genes."

35 Rita Arditti, "Bobby Seale at Cold Spring Harbor," *Science for the People* 2, (August 1970): 3.

36 James Watson, remarks, annual report, Cold Spring Harbor Laboratory (1969), Cold Spring Harbor Laboratory Library and Archives.

37 Prior to *Science for the People*, there had been seven issues of the national SESPA newsletter, but this eighth issue took up the slogan and logo from the buttons at the December 1969 Boston AAAS conference.

38 Editorial Collective, "Equality for Women in Science," *Science for the People* 2 (August 1970): 10–11.

39 N. C. Khanduri in the SESPA newsletter, quoted in Editorial Collective, "Equality for Women in Science."

40 Kuljian, interview with Muchnik.

41 In December 1971, at its annual meeting in Philadelphia, the AAAS held a session titled "Women in Science" at which a women's caucus was formed. The caucus drafted a proposal to establish an "office for women's equality." As a result, AAAS established its Office of Opportunities in Science, which would address the needs of both women and people of color. In December 1972, Janet Welsh Brown was appointed director, and she would go on, with Paula Quick Hall and Shirley Malcom, to author "The Double Bind" report in 1976. See Margaret Rossiter, *Women Scientists in America*, vol. 3, *Forging a New World Since 1972* (Baltimore: Johns Hopkins University, 2012), 8.

42 Patricia Mack Whitaker-Azmitia, "The Discovery of Serotonin and Its Role in Neuroscience," *Neuropsychopharmacology* 21 (1999): 2–8.

43 Christa Kuljian, interview with Ursula Goodenough, November 30, 2019. In 1972, Carolyn Cohen moved her cancer lab to Brandeis University, where she worked for another forty years, until she retired in 2012 ("Retiring Faculty Members Honored at Luncheon," *Brandeis Now*, March 13, 2024).

44 Sara Corbett, "The Lives They Lived: Ruth Hubbard Challenged the Male Model of Science," *New York Times Magazine*, December 21, 2016.

45 Ursula Goodenough, comment, in "In Celebration of Ruth Hubbard," symposium, June 2, 1990, Ruth Hubbard Papers, box 13, folder 6, Schlesinger Library, Harvard University.

46 Walzer, interview, 106.

47 Walzer, interview, 106.

48 I'm grateful to Elijah Wald for sharing his mother's summer 1970 journal with me.

49 Spanier, interview with Hubbard, c. 1986.

50 Kuljian, interview with Muchnik; Christa Kuljian, interview with Estelle Disch, September 24, 2019.

51 Evelyn F. Keller and Lee A. Segel, "Initiation of Slime Mold Aggregation Viewed as an Instability," *Journal of Theoretical Biology* 26, no. 3 (1970): 399–415; Keller, *Making Sense of My Life in Science*, 78. Thanks to Nadja Cech for discussing this paper with me.

52 Evelyn Fox Keller, "From Working Scientist to Feminist Critic," in *The Gender and Science Reader*, ed. Muriel Lederman and Ingrid Bartsch (London: Routledge, 2001), 60; Keller, "Women in Science: An Analysis of a Social Problem," *Harvard Magazine* 77, no. 2 (1974): 16.

53 Keller, Making Sense of My Life in Science, 81–82; Keller, "From Working Scientist to Feminist Critic," 60.

54 Christina Agapakis, "Conversations with Evelyn Fox Keller," *Method Quarterly* (November 2014), http://www.methodquarterly.com; Keller, *Making Sense of My Life in Science*, 81.

CHAPTER 4

1 Bonnie Spanier, interview with Rita Arditti, c. 1978, box 1, folder 6, Bonnie Spanier Papers, Schlesinger Library, Harvard University.

2 Not long after serving on the commission, Colleen Meyer died by suicide. Arditti believed that her experience on the commission had contributed to her death (Spanier, interview with Arditti, c. 1978).

3 Rita Arditti and Tom Strunk, "Objecting to Objectivity: A Course in Biology," *Science for the People* 4, no. 5 (1972): 16–21.

4 Arditti and Strunk, "Objecting to Objectivity."

5 Arditti and Strunk, "Objecting to Objectivity," 18.

6 Carol Axelrod and Ruth Crocker, "The Natural Birth of a Women's Group," *Science for the People* 6, no. 5 (1974): 15.

7 Axelrod and Crocker, "The Natural Birth of a Women's Group," 15.

8 Axelrod and Crocker, "The Natural Birth of a Women's Group," 15.

9 Rita Arditti, "Women's Biology in a Man's World: Some Issues and Questions," *Science for the People* 5, no. 4 (1973): 39–42.

10 Arditti, "Women's Biology in a Man's World," 39.

11 Arditti, "Women's Biology in a Man's World," 42.

12 "Reports of the Project Groups: Women's Issue Project," *Science for the People* 4, no. 6 (1972): 33.

13 Unnamed author [likely Freda Salzman and other members of the women's issues group of Science for the People], unpublished draft of an introduction to *Science for the People*'s special issue on women, box 6, folder 155, Freda Salzman Papers, Schlesinger Library, Harvard University; Arthur Jensen, "How Much Can We Boost IQ and Scholastic Achievement?" *Harvard Educational Review* 39, no. 1 (1969): 11–23.

14 Richard Herrnstein, "IQ," *Atlantic Monthly* (September 1971): 43–64; Peter Shapiro, "A Spring of Rekindled Activism," *Harvard Crimson*, September 1, 1972.

15 By 1973, the Black Panther Party had spoken out against Jensen's article and scientific methodology. The party did not oppose scientific research but pointed out that "few scientists seem interested in comparing say Northern Italians with Southern Italians, . . . Appalachian Whites with Social Register Whites." See Alondra Nelson's excellent *Body and Soul: The Black Panther Party and the Fight against Medical Discrimination*, (Minneapolis: University of Minnesota Press, 2011), esp. 165–67.

16 Unnamed author, unpublished draft of an introduction.

17 Spanier, interview with Arditti, c. 1978.

18 Spanier, interview with Arditti, c, 1978.

19 Daphne Spaine, "Women's Rights and Gendered Spaces in 1970s Boston," *Frontiers* 32, no. 1 (2011): 168.

20 Rita Arditti, "Women as Objects: Science and Sexual Politics," *Science for the People* 6, no. 5 (1974): 8.

21 Arditti, "Women as Objects," 8.

22 Arditti, "Women as Objects," 8.

23 Arditti, "Women as Objects," 9.

24 Arditti, "Women as Objects," 30–31.

25 Christa Kuljian, interview with Ursula Goodenough, November 30, 2019.

26 Ann Juergens, "The Status of Women: Is Harvard Progressing?" *Harvard Crimson*, June 15, 1972.

27 "Ruth Hubbard," in *How to Think about Science* (Ottawa: Canadian Broadcasting Corporation, 2009); Judith Walzer, interview with Ruth Hubbard (1981), Judith Walzer Papers, Schlesinger Library, Harvard University, 109.

28 Walzer, interview, 110.

29 "Ruth Hubbard"; Walzer, interview, 112.

30 Ruth Hubbard, "Reflections on my Life as a Scientist," *Radical Teacher* 30 (January 1986): 6.

31 Ruth Hubbard, letter to Margaret Randall, September 12, 1987, in *The Shape of Red: Insider/Outsider Reflections*, ed. Ruth Hubbard and Margaret Randall (San Francisco: Cleis, 1988), 161; Dolita Cathcart, personal communication, March 9, 2019; Deborah Wald, personal communication, September 21, 2023.

32 Ruth Hubbard, "Remarks at Macalester College Commencement" (May 1991), box 14, folder 4, Ruth Hubbard Papers, Schlesinger Library, Harvard University.

33 Hubbard, "Remarks at Macalester College Commencement." I am grateful to Elijah Wald and Deborah Wald for describing the influence of the Black Panther Party on Hubbard's awareness: personal communications, September 21, 2023.

34 Stephen J. Gould and John Berg, "Academic Racism," *Harvard Crimson*, April 30, 1974.

35 Walzer, interview, 115.

36 Walzer, interview, 116.

37 Walzer, interview, 6, 15–20.

38 Margaret Rossiter, *Women Scientists in America*, vol. 2, *Forging a New World Since 1972* (Baltimore: Johns Hopkins University Press, 2012), 21–25.

39 Hubbard, "Reflections on my Life as a Scientist."

40 Hubbard, "Reflections on my Life as a Scientist."

41 Christa Kuljian, interview with Elijah Wald, April 23, 2020.

42 Ruth Hubbard, untitled poem, March 20, 1974, Ruth Hubbard Papers, Schlesinger Library, Harvard University.

43 Walzer, interview, 114.

44 Mary Sue Henifin, video comments, in "In Celebration of Ruth Hubbard," symposium, June 2, 1990, Vt-232–0001, Ruth Hubbard Papers, Schlesinger Library, Harvard University.

45 Walzer, interview, 123–25.

46 Gay Seidman, "When Cooler Heads Don't Prevail," *Harvard Crimson*, November 6, 1976.

47 Walzer, interview, 120.

48 Deborah Wald, comment, in "In Celebration of Ruth Hubbard."

49 Catherine Russo, video interview with Rita Arditti (1991), https://ritaarditti.com.

50 Rita Arditti, with Estelle Disch, "Not Dead Yet: Living with Stage IV Breast Cancer" (2016), https://ritaarditti.com.

51 Arditti, "Not Dead Yet."

52 Arditti, "Not Dead Yet."

53 Kuljian, interview with Muchnik.

54 Christa Kuljian, interview with Deborah Wald, April 23, 2020.

55 Kuljian, interview with Muchnik; Muchnik, personal communication.

CHAPTER 5

1 Bonnie Spanier, audio interview with Evelynn Hammonds, c. 1986, Bonnie Spanier Papers, Schlesinger Library, Harvard University. Hammonds mentions the election incident in Aimee Sands, "Never Meant to Survive: A Black Woman's Journey: An Interview with Evelynn Hammonds," in *The Racial Economy of Science: Toward a Democratic Future*, ed. Sandra Harding (Bloomington: Indiana University Press, 1993), 241. Also Christa Kuljian, interviews with Evelynn Hammonds, August 18, 2020, and November 19, 2021.

2 Spanier, interview with Hammonds.

3 Spanier, interview with Hammonds; Sands, "Never Meant to Survive," 241; Kuljian, interviews with Hammonds, August 18, 2020, and November 19, 2021.

4 Spanier, interview with Hammonds; Sands, "Never Meant to Survive," 241. Kuljian, interviews with Hammonds, August 18, 2020, and November 19, 2021.

5 In 1973, fewer than fifty women, compared to more than 1,200 men, completed a PhD in physics in the United States (American Institute of Physics, https://www.aip.org).

6 Evelynn Hammonds, speech, presentation of W. E. B. DuBois Award to Shirley Jackson, October 11, 2018, https://www.youtube.com. In 1995 Jackson would go on to become chair of the Nuclear Regulatory Commission, and in 1999 she would become president of Rensselaer Polytechnic Institute, a position she would hold for twenty-three years.

7 Sands, "Never Meant to Survive," 241.

8 Spanier, interview with Hammonds.

9 Christa Kuljian, interview with Hammonds, August 18, 2020.

10 Spanier, interview with Hammonds.

11 Sands, "Never Meant to Survive," 239; Spanier, interview with Hammonds.

12 Sands, "Never Meant to Survive," 241; Kuljian, interview with Hammonds, August 18, 2020.

13 Clarence G. Williams, "Evelynn M. Hammonds," in *Technology and the Dream: Reflections on the Black Experience at MIT, 1941–1999* (Cambridge, MA: MIT Press, 2001), 937–44; Spanier, interview with Hammonds.

14 Spanier, interview with Hammonds.

15 Spanier, interview with Hammonds; Sands, "Never Meant to Survive," 243; Kuljian, interview with Hammonds, August 18, 2020.

16 Spanier, interview with Hammonds; Kuljian, interview with Hammonds, August 18, 2020.

17 Spanier, interview with Hammonds; Kuljian, interview with Hammonds, August 18, 2020.

18 In December 1972, Janet Welsh Brown, a political scientist, became the first director of the Office of Opportunities in Science at AAAS, a position she held until April 1, 1979, when she left to head the Environmental Defense Fund. Paula Quick Hall trained as a psychologist and later earned her doctorate in political science, specializing in public administration and public policy. On Shirley Malcom, see *The History Makers: The Digital Repository for the Black Experience*, https://www.thehistorymakers.org.

19 Hammonds has often spoken of this letter and Malcom's response. See, for instance, their discussion during "The Roundtable on Black Men and Black Women in Science, Engineering, and Medicine in Conversation with Living Legend Dr. Shirley Malcom," hosted by the National Academies of Sciences, Engineering, and Medicine, February 24, 2022, http://www.nationalacademies.org.

20 Spanier, interview with Hammonds.

21 Spanier, interview with Hammonds.

22 Shirley Jackson was instrumental in increasing the numbers of Black students at MIT. After the assassination of Martin Luther King Jr. in 1968, she helped to found the Black Students Union and began to push for greater recruitment of both students and faculty of color. In 1969, fifty-seven African American undergraduates matriculated, more than ten times the number in any other year in MIT's history (Hammonds, speech, presentation of W. E. B. DuBois Award). In 1949, James Earl Young graduated with a bachelor's degree in physics from Howard University, and he earned his PhD in physics at MIT in 1953. In 1970, he became the first Black tenured professor in the physics department. In 1977, he was a founding member of the National Society of Black Physicists. Young held his position at MIT until he retired in 1992 (https://physics.mit.edu).

23 Spanier, interview with Hammonds; Kuljian, interview with Hammonds, August 18, 2020.

24 Yosano Akiko's poem, whose title translates as "Mountain-Moving Day," was first published in the Japanese journal *Seito* (September 1911). In 1972, the Chicago Women's Graphic Collective translated this excerpt and created the poster (https://collections.museumca.org).

25 Spanier, interview with Hammonds; Kuljian, interview with Hammonds, August 18, 2020.

26 Boston Women's Health Book Collective, *Our Bodies, Ourselves* (New York: Simon and Schuster: 1976).

27 Hammonds, speech, "Radical Roots" panel, Black Feminist Health Science Studies Symposium, Massachusetts Institute of Technology, March 24, 2021, https://www.youtube.com.

28 Williams, *Technology and the Dream*, 938.

29 Williams, *Technology and the Dream*, 939.

30 Evelynn Hammonds, "Underrepresentations," *Science*, August 23, 1991, 919.

31 Spanier, interview with Hammonds.

32 Evelynn Hammonds, "X-Ray Scattering of Xenon Adsorbed on Graphite" (master's thesis, Massachusetts Institute of Technology, 1980); E. M. Hammonds, P. Heiney, P. W. Stephens, R. J. Birgeneau, and P. Horn, "Structure of Liquid and Solid Monolayer Xenon on Graphite," *Journal of Physics* C 13, no. 12 (1980) L301.

33 Sands, "Never Meant to Survive," 247.

34 Sands, "Never Meant to Survive," 248.

35 Evelyn Fox Keller, "From Working Scientist to Feminist Critic," in *The Gender and Science Reader*, ed. Muriel Lederman and Ingrid Bartsch (New York: Routledge, 2001), 60.

36 Evelyn Fox Keller, "Women in Science," *Harvard Magazine* 77, no. 2 (1974): 16.

37 Keller, "From Working Scientist to Feminist Critic," 60; Christina Agapakis, "Conversations with Evelyn Fox Keller" (November 2011), *Method Quarterly*, http://www.methodquarterly.com.

38 Keller, "Women in Science," 17.

39 Keller, "Women in Science," 17.

40 Keller, "Women in Science," 18.

41 Keller, "Women in Science," 18.

42 Keller, "From Working Scientist to Feminist Critic," 61.

43 Agapakis, "Conversations with Evelyn Fox Keller."

44 Keller, "The Anomaly of a Woman in Physics," in *Women, Science, and Technology: A Reader in Feminist Science Studies*, ed. Mary Wyer, Mary Barbercheck, Donna Geisman, Hatice Orun Ozturk, and Marta Wayne (New York: Routledge, 2001),16.

45 Margaret Rossiter, speech, "1970–2000: A Less Than Golden Age for Women in Chemistry," Cornell University (2000), https://www.ncbi.nlm.nih.gov; Susan Dominus, "Women Scientists Were Written Out of History. It's Margaret Rossiter's Lifelong Mission to Fix That," *Smithsonian* 50, no. 6 (2019): 42–53.

46 Margaret Rossiter, *Women Scientists in America*, vol. 1, *Struggles and Strategies to 1940* (Baltimore: Johns Hopkins University Press, 1982), xi.

47 Margaret Rossiter, "Women Scientists in America before 1920," *American Scientist* 62, no. 3 (1974): 312–23.

CHAPTER 6

1 E. O. Wilson, *Sociobiology: The New Synthesis* (1975; reprint, Cambridge, MA: Belknap Press, 2000), 4.

2 Boyce Rensberger, "Sociobiology: Updating Darwin on Behavior," *New York Times*, May 28, 1975.

3 Wilson, *Sociobiology: The New Synthesis*, 550.

4 Jon Beckwith, *Making Genes, Making Waves* (Cambridge, MA: Harvard University Press, 2002), 139.

5 Ullika Segerstrale, *Defenders of the Truth* (Oxford: Oxford University Press, 2000), 19.

6 Elizabeth Allen, Barbara Beckwith, Jon Beckwith, Steven Chorover, David Culver, et al., "Against 'Sociobiology,'" *New York Review of Books*, November 13, 1975.

7 Wilson, *Sociobiology*, 553.

8 Hubbard critiqued sociobiology for decades. See, for instance, Ruth Hubbard, "Sexism and Sociobiology" *Psychology Today* (1978); Ruth Hubbard, "Do Only Men Evolve?," in *Women Look at Biology Looking at Women* (Cambridge, MA: Schenkman, 1979), 7–36; Ruth Hubbard and Marian Lowe, "Sociobiology and Biosociology: Can Science Prove the Biological Basis of Sex Differences in Behavior?," in *Genes and Gender II*, ed. Ruth Hubbard and Marian Lowe (New York: Gordian, 1979), 91–111; Ruth Hubbard, "Facts and Feminism," *Science for the People* 18, no. 2 (1986):16–20; and Ruth Hubbard, *The Politics of Women's Biology* (New Brunswick, NJ: Rutgers University Press, 1990), 107–18.

9 Beckwith, *Making Genes*, 140.

10 C. H. Waddington, who wrote the original *New York Review of Books* article on Wilson's work, did not respond to the letter because he died on September 26, 1975. See E. O. Wilson, "For Sociobiology," *New York Review of Books*, December 11, 1975.

11 Sarag Shang, "Harvard's 'Frankenstein': The 70s Controversy over Mixing DNA," March 30, 2015, https://gizmodo.com.

12 E. O. Wilson, "Foreword," in *The Sociobiology Debate*, ed A. Caplan (New York: Harper and Row, 1978); E. O. Wilson, *Naturalist* (Washington, DC: Island Press, 1994), 339.

13 Christa Kuljian, interview with Marian Lowe, March 18, 2021.

14 Sandra Korn, "Sex, Science, and Politics in the Sociobiology Debate" (PhD diss., Harvard University, 2014), 78; Desmond Morris, *The Naked Ape* (New York: McGraw-Hill, 1967).

15 Margaret Rossiter, *Women Scientists in America*, vol. 2, *Before Affirmative Action, 1940–1972* (Baltimore: Johns Hopkins University Press, 1995), 122–48; Kuljian, interview with

Lowe. On how the nepotism law shaped Freda Salzman's career, see Maggie Chen, "Wives, Physics, and Nepotism in Academica," *Lady Science*, August 18, 2020, https://www.lady science.com.

16 Wilson, *Sociobiology*, 547.

17 Doris O'Donnell, letter to E. O. Wilson, May 27, 1976, box 4, folder 119, Freda Salzman Papers, Schlesinger Library, Harvard University.

18 Anne Fausto-Sterling, review of *Defenders of the Truth*, *Genome News Network*, January 16, 2001.

19 Wilson, *Sociobiology*, 555.

20 E. O. Wilson's papers include an entire folder of correspondence with Robert Ardrey (Nayanika Ghosh, personal communication, December 1, 2022). Ghosh will soon publish her analysis of this correspondence.

21 E. O. Wilson, *On Human Nature* (Cambridge, MA: Harvard University Press, 1978), 128–33.

22 Jon Beckwith, "The Radical Science Movement in the United States," *Monthly Review* 38, no. 3 (1990): 123.

23 Sociobiology Study Group, meeting minutes, July 27, 1976, box 3, folder 87, Freda Salzman Papers, Schlesinger Library, Harvard University. Also see Kelly Moore, *Disrupting Science* (Princeton, NJ: Princeton University Press, 2008), 184.

24 Segerstrale, *Defenders of the Truth*, 222.

25 Richard Lewontin, letter to the editor, *Bioscience* 29 (June 1979): 341–44.

26 Wilson, *Sociobiology*, 317, 567–68; Sociobiology Study Group, meeting minutes, December 7, 1976, box 4, folder 118, Freda Salzman Papers, Schlesinger Library, Harvard University.

27 Korn, "Sex, Science, and Politics in the Sociobiology Debate," 73. Korn kindly sent me the entire film in November 2020.

28 In 1972, DeVore, Trivers and Wilson were interviewed for a Canadian television series but never heard from the production crew again. They assumed the project had been canceled. Then, in 1976, they received a brochure for the film, now titled *Sociobiology: Doing What Comes Naturally*. They saw the film for the first time at the Science for the People showing in December of that year and were dismayed. "The three of us saw the film with some shock," said DeVore, "not for what we said in the interviews—much of which we still stand by, but because of the visuals and the music used as background" (Roger Lewin, "Sociobiologists Cry 'Foul' over New Film," *New Scientist*, December 15, 1977, 711). Also see Irven DeVore, Robert Trivers, and E. O. Wilson, letter to *Science for the People*, March 30, 1977, box 3, folder 87, Freda Salzman Papers, Schlesinger Library, Harvard University; and Cora Stuhrmann, "Sociobiology on Screen: The Controversy through the Lens of 'Sociobiology: Doing What Comes Naturally,'" *Journal of the History of Biology*, June 29, 2023, 52.

29 Moore, *Disrupting Science*, 184.

30 Glenn Wargo and Scott Schneider, "Voluntown, 1977—SftP Conference," *Science for the People* 9, no. 4 (1977): 33–37.

31 Barbara Chasin, "Sociobiology: A Sexist Synthesis," *Science for the People* 9, no. 2 (1977): 27.

32 "Are Sex Roles Biologically Determined?," conference program, March 15, 1977, box 4, folder 118, Freda Salzman Papers, Schlesinger Library, Harvard University.

33 Freda Salzman, "Are Sex Roles Biologically Determined?," *Science for the People* 9, no. 4 (1977): 27–32.

34 Lila Leibowitz, letter to Ruth Hubbard, unidentified date, Lila Leibowitz Papers, Schlesinger Library, Harvard University.

35 "Why You Do What You Do: Sociobiology: A New Theory of Behavior," *Time*, August 1, 1977, https://content.time.com. Trivers's work was based on the mathematics of the British biologist William Hamilton, done in 1964. Trivers wrote about reciprocal altruism in 1971, and E. O. Wilson brought these ideas together in 1975.

CHAPTER 7

1 Ruth Hubbard, Mary Sue Henifin, and Barbara Fried, "Introduction," in *Women Look at Biology Looking at Women: A Collection of Feminist Critiques* (Cambridge, MA: Schenkman, 1979), xvii.

2 Hubbard et al., "Introduction," xvii.

3 Hubbard et al., "Introduction," xix–xx.

4 Hubbard et al., "Introduction," xx.

5 Ruth Hubbard, "Have Only Men Evolved?," in *Women Look at Biology Looking at Women*, 7.

6 Hubbard, "Have Only Men Evolved?," 11–12.

7 Hubbard, "Have Only Men Evolved?," 12.

8 Hubbard, "Have Only Men Evolved?," 12.

9 Hubbard, "Have Only Men Evolved?," 13.

10 Hubbard, "Have Only Men Evolved?," 16.

11 Hubbard, "Have Only Men Evolved?," 16

12 Hubbard, "Have Only Men Evolved?," 17

13 Hubbard, "Have Only Men Evolved?," 17.

14 Hubbard, "Have Only Men Evolved?," 18–19.

15 Hubbard, "Have Only Men Evolved?," 19.

16 Hubbard, "Have Only Men Evolved?," 19.

17 Hubbard, "Have Only Men Evolved?," 19–20.

18 Hubbard, "Have Only Men Evolved?," 16; Rebekkah Rubin, "The Woman Who Challenged Darwin's Sexism," *Smithsonian Magazine*, November 9, 2017, http://smithsonianmag.com.

19 Hubbard, "Have Only Men Evolved?," 21.

20 Hubbard, "Have Only Men Evolved?," 21.

21 E. O. Wilson, *Sociobiology: The New Synthesis* (1975; reprint, Cambridge, MA: Belknap Press, 2000), 316.

22 Hubbard, "Have Only Men Evolved?," 24–25.

23 Hubbard, "Have Only Men Evolved?," 26.

24 Hubbard, "Have Only Men Evolved?," 29.

25 Hubbard, "Have Only Men Evolved?," 31.

26 Hubbard, "Have Only Men Evolved?," 32.

27 Ruth Hubbard, Mary Sue Henifin, and Barbara Fried, "Epilogue," in *Women Look at Biology Looking at Women*, 206.

28 Hubbard et al., "Epilogue," 208

29 Hubbard et al., "Epilogue," 208.

30 Garland Allen, letter to the editor, *Science for the People* 8, no. 6 (1976): 27.

31 Allen, letter to the editor, 27.

32 Rita Arditti, letter to the editor, *Science for the People* 9, no. 1 (1977): 5.

33 Sadly, I was not able to speak to Allen before his death. See Jane Maienshein, "Garland E Allen III: In Memoriam (1936–2023)," *Science* 379, no. 6639 (2023): 1304.

34 Ruth Hubbard, letter to the editor, *Science for the People* 9, no. 1 (1977): 5.

35 Lorraine Roth, letter to the editor, *Science for the People* 9, no. 2 (1977): 5.

36 Christa Kuljian, interview with Jonathan Beckwith, June 15, 2019.

CHAPTER 8

1 Ethel Tobach and Betty Rosoff, *Genes and Gender: On Hereditarianism and Women* (New York: Gordian, 1978), 7.

2 Dorothy Burnham, "Biology and Gender," in *Genes and Gender,* 51.

3 Burnham, "Biology and Gender," 52.

4 Burnham was quoting Truth accurately as of 1977, but later transcriptions vary. See https://voicesofdemocracy.umd.edu.

5 Burnham, "Biology and Gender," 54.

6 Burnham, "Biology and Gender," 58.

7 Burnham, "Biology and Gender," 58.

8 Tobach and Rosoff, *Genes and Gender,* 7.

9 The proceedings of the AAAS symposium were later published, but they did not include the panel organized by Hubbard or the others organized by Science for the People. Also see George W. Barlow and James Silverberg, eds., *Sociobiology: Beyond Nature/Nurture?* (New York: Routledge, 1980).

10 Ruth Hubbard and Marian Lowe, eds., *Genes and Gender II: Pitfalls in Research on Sex and Gender* (New York: Gordian, 1979).

11 This episode has been retold many times. See Ullica Segerstrale, *Defenders of the Truth: The Battle for Science in the Sociobiology Debate and Beyond* (Oxford: Oxford University Press, 2000), 23; Jon Beckwith and Bob Lange, "AAAS: Sociobiology on the Run," *Science for the People* (March–April 1978), 38–39; and Erika Lorraine Milam, *Creatures of Cain: The Hunt for Human Nature in Cold War America* (Princeton, NJ: Princeton University Press, 2019), 259.

12 Segerstrale, *Defenders of the Truth,* 23; Beckwith and Lange, "AAAS: Sociobiology on the Run," 38–39; Milam, *Creatures of Cain,* 259.

13 Later *Genes and Gender* volumes focused on *sociobiology and the implications for children* (Ethel Tobach and Betty Rosoff, eds., *Genes and Gender III: Genetic Determinism and Children* [New York: Gordian, 1980]); heredity and women's health (Ethel Tobach and Betty Rosoff, eds., *Genes and Gender IV: The Second X and Women's Health* [New York: Gordian, 1983]); socialization toward inequality (Ethel Tobach and Betty Rosoff, eds., *Genes and Gender V: Women at Work* [New York, Gordian, 1988]); and the speciousness of genetic explanations (Ethel Tobach and Betty Rosoff, eds., *Genes and Gender VI: On Peace, War, and Gender* [New York: Gordian, 1991]). The final volume in the series was the first to directly address racism and included an excellent essay by Dorothy Burnham's daughter Linda Burnham, the cofounder and director of the Women of Color Resource Center in Oakland, California (Ethel Tobach and Betty Rosoff, eds., *Genes and Gender VII: Challenging Racism and Sexism* [New York: Feminist Press, 1994]). For a thorough discussion, see Sandra Korn, "Sex, Science, and Politics in the Sociobiology Debate," (PhD diss., Harvard University, 2014), 114–21.

14 Nancy M. Tooney, "Preliminary Meeting of February 3, 1979," meeting notes for "Beyond

Body and Behavior Workshop/Conference," box 2, folder 47, Lila Leibowitz Papers, Schlesinger Library, Harvard University. Sandra Korn pointed me toward this archive, and I am grateful.

15 Ruth Bleier, letter to Marian Lowe, March 7, 1979, box 2, folder 58, Lila Leibowitz Papers, Schlesinger Library, Harvard University.

16 Bleier, letter to Lowe.

17 Ruth Hubbard, Marian Lowe, Lila Leibowitz, and Mary Howell, letter addressed to "Dear Colleague," January 2, 1979, box 2, folder 47, Lila Leibowitz Papers, Schlesinger Library, Harvard University; Tooney, "Preliminary Meeting of February 3, 1979."

18 Boston Genes and Gender Group, letter addressed to "Dear Companeras!," June 13, 1979, box 2, folder 47, Lila Leibowitz Papers, Schlesinger Library, Harvard University.

19 Marian Lowe, document with headings "Friday Night Session," "Saturday Sessions," and "Sunday Sessions," box 2, folder 47, Lila Leibowitz Papers, Schlesinger Library, Harvard University.

20 Ruth Hubbard, letters to Boston Genes and Gender Group, July 8, 1979, and July 27, 1979; and Evelynn Hammonds, Michelle Harrison, Mary Howell, Ruth Hubbard, Pat Hynes, Marian Lowe, Jan Raymond, and Susan Yakutis, memo to participants in the "Beyond Determinist Thinking" conference, July 27, 1979; all in box 2, folder 58, Lila Leibowitz Papers, Schlesinger Library, Harvard University. Regarding her custard question, Hubbard concluded that it has no answer. Depending on many factors, custard may set, liquefy, curdle, or taste terrible.

21 Ruth Bleier, memo to participants in the "Beyond Determinist Thinking" conference, August 13, 1979, box 2, folder 58, Lila Leibowitz Papers, Schlesinger Library, Harvard University.

22 National Women's Studies Association, conference program, May 30–June 3, 1979, https://drum.lib.umd.edu.

23 The Boston Genes and Gender Group did not meet again after the "Beyond Determinist Thinking" conference. Several participants were disappointed with how it turned out, especially in relation to a debate about "hetero-sexist patriarchy" (Mary Howell, letter to participants, September 18, 1979, box 2, folder 47, Lila Leibowitz Papers, Schlesinger Library, Harvard University; Christa Kuljian, interview with Anne Fausto-Sterling, January 26, 2021).

24 Christa Kuljian, interview with Evelynn Hammonds, August 18, 2020.

25 Nancy Krieger, letter to Ruth Hubbard, July 21, 1979, box 2, folder 17, Ruth Hubbard Papers, Schlesinger Library, Harvard University.

CHAPTER 9

1 For the full story of the founding of the Cambridge Women's Center, see the film *Left on Pearl*, directed by Susan Rivo (2016).

2 Beverly Smith's field placement was sponsored by the Boston Family Planning Project, which had federal funding to provide reproductive health services at various sites around the region, including Boston City Hospital.

3 Susan Goodwillie [Stedman], interview with Barbara Smith, 1994, quoted in Duchess Harris, "'All of Who I Am in the Same Place': The Combahee River Collective," in *Womanist Theory and Research* 2, no. 1 (1999): 1. Despite due diligence, I have not been able to fully trace the Stedman source.

4 Keeanga-Yamahtta Taylor, ed., *How We Get Free* (Chicago: Haymarket, 2017), 98; Christa Kuljian, interview with Beverly Smith, November 27, 2020; Beverly Smith, personal communication, November 6, 2023. On the Women's Community Health Center, see Daphne Spain, "Women's Rights and Gendered Spaces in 1970s Boston," *Frontiers* 32, no. 1 (2011): 158–61.

5 Kuljian, interview with Beverly Smith.

6 Susan Goodwillie [Stedman], interview with Demita Frazier, 1994, quoted in Harris, "All of Who I Am in the Same Place," 1; Taylor, *How We Get Free*, 54, 126. Despite due diligence, I have not been able to fully trace the Stedman source.

7 Barbara Smith, "What Would Harriet Do? A Legacy of Resistance and Activism: A Conversation with Barbara Smith and Beverly Guy-Sheftall," *Meridians* 12, no. 2 (2014): 127–28; Taylor, *How We Get Free*, 126–27.

8 Taylor, *How We Get Free*, 101.

9 Taylor, *How We Get Free*, 44–45.

10 Harris, "All of Who I Am in the Same Place."

11 Beverly Smith, personal communication, November 6, 2023. On November 17, 2023, Barbara Smith tweeted, "We wrote the Combahee River Collective Statement in 1977 when Zillah Eisenstein invited us to contribute to her book." Also see Taylor, *How We Get Free*, 61, 123.

12 Zillah R. Eisenstein, ed., *Capitalist Patriarchy and the Case for Feminist Socialism* (New York: Monthly Review Press, 1978).

13 Taylor, *How We Get Free*, 19.

14 Taylor, *How We Get Free*, 26–27.

15 Taylor, *How We Get Free*, 21.

16 Taylor, *How We Get Free*, 23.

17 Kuljian, interview with Beverly Smith.

18 Barbara Smith, "Notes for Yet Another Paper on Black Feminism, or Will the Real Enemy Please Stand Up?," *Conditions* 5, no. 2 (1979): 124.

19 Taylor, *How We Get Free*, 58.

20 Christa Kuljian, interview with Evelynn Hammonds, August 18, 2020.

21 Taylor, *How We Get Free*, 106; Beverly Smith, personal communication, November 6, 2023.

22 "Two Bodies Found in a Trash Bag," *Boston Globe*, January 30, 1979; Dave Wood, "Dorchester Girls Found Dead," *Boston Globe*, January 31, 1979; Mike Barnicle, "When Tragedy Isn't News," *Boston Globe*, February 1, 1979; Timothy Dwyer, "Fourth Black Woman Found Slain in Boston," *Boston Globe*, February 6, 1979; Isabel Lonnie and Dave Wood, "Fifth Black Woman Slain in South End," *Boston Globe*, February 22, 1979; Chris Black, "Slain Woman Student, 17, of Dorchester," *Boston Globe*, March 16, 1979; Gayle Pollard, Carmen Fields, and Viola Osgood, "Six Slain Women, and Those Who Loved Them," *Boston Globe*, April 1, 1979. The names of the six women gleaned from the press coverage were Christine Ricketts, Andrea Foye, Gwendolyn (Yvette) Stintson, Caron Prater, Daryal Ann Hargett, and Desiree Etheridge. For a thorough review of the murders and the reaction to them, see Jaime M. Grant, "Who's Killing Us?" *Sojourner* 10 (June 1988): 20–21.

23 Beverly Smith, personal communication, November 7, 2023.

24 Grant, "Who's Killing Us?," 21.

25 Combahee River Collective, "Six Black Women: Why Did They Die?," *Aegis*, May 1, 1979.

26 Jaime M. Grant, "Who's Killing Us?: Part II," *Sojourner* 11 (1988): 16–18. Also see Dialynn Dwyer, "11 Black Women Were Murdered in Boston 40 Years Ago. A Local Artist Is Remembering Them across the City," February 20, 2019, https://www.boston.com.

27 Alexis De Veaux, *Warrior Poet: A Biography of Audre Lorde* (New York: Norton, 2004), 242–43.

28 Bonnie Spanier, "How I Came to This Study," in *The Gender and Science Reader*, ed. Muriel Lederman and Ingrid Bartsch (New York: Routledge, 2000), 55. In this essay, Spanier recalls the impact of the murders in 1979 and how they affected her as a scientist in Hubbard's class. Also Christa Kuljian, interview with Bonnie Spanier, March 30, 2021.

29 Beverly Smith, note to Ruth Hubbard, August 11, 1981, Ruth Hubbard Papers, Harvard University Archives.

30 Ruth Hubbard, Mary Sue Henifin, and Barbara Fried, eds., *Biological Woman—The Convenient Myth* (Cambridge, MA: Schenkman, 1982), xiii.

31 Helen Rodriguez-Trias, "Sterilization Abuse," in *Biological Woman*, 148, 149.

32 Beverly Smith, "Black Women's Health: Notes for a Course," in *Biological Woman*, 227.

33 Boston Women's Health Book Collective, *The New Our Bodies, Ourselves: The Fifth Edition* (New York: Simon and Schuster, 1984), xiii.

34 For reactions to Lorde's speech, see Michelle Moravec, "Unghosting Apparitional (Lesbian) History: Erasures of Black Lesbian Feminism, 11 January 2019," https://scalar.usc .edu. Moravec wonders, "Did Lorde refrain from recommending women from the [Combahee River] Collective as participants in the conference?" (ibid.). Also *see* Carol Anne Douglas and Brooke Douglas, "2nd Sex 30 Years Later: Feminist Theory Conference," *Off Our Backs* 4 (December 1979): 24–27; Lester C. Olson, "The Personal, the Political, and Others: Audre Lorde Denouncing the Second Sex Conference," *Philosophy and Rhetoric* 33, no. 3 (2000): 259–85; and Jessica Benjamin, "Letter to Lester Olson," *Philosophy and Rhetoric* 33, no. 3, (2000): 286–90.

35 Audre Lorde, "The Master's Tools Will Never Dismantle the Master's House," in *Sister Outsider: Essays and Speeches* (1984; reprint, Berkeley, CA: Crossing Press, 2007), 110.

36 Lorde, "The Master's Tools," 110.

37 Lorde, "The Master's Tools," 112–13.

CHAPTER 10

1 Evelyn Fox Keller, *Making Sense of My Life in Science: A Memoir* (Amherst, MA: Modern Memoirs, 2023), 110.

2 Ullica Segerstrale, *Defenders of the Truth: The Battle for Science in the Sociobiology Debate and Beyond* (Oxford: Oxford University Press, 2000), 21.

3 Keller, *Making Sense of My Life in Science*, 123.

4 Eveyln Fox Keller, "Gender and Science," in *Reflections on Gender and Science* (New Haven, CT: Yale University Press, 1985), 75–76.

5 Keller, "Gender and Science."

6 Keller, "Gender and Science," 77.

7 Evelyn Fox Keller, "Nature as 'Her,'" unpublished conference paper, n.d., box 3, folder 97, Freda Salzman Papers, Schlesinger Library, Harvard University. I'm not sure if Salzman attended "The Second Sex" conference, if the papers were distributed ahead of time, or if Keller sent a copy to her directly.

8 Christa Kuljian, interview with Evelyn Fox Keller, September 17, 2019; Keller, *Making Sense of My Life in Science*, 107.

9 Christa Kuljian, interviews with Evelyn Fox Keller, September 17, 2019, and September 27, 2019. 2019; Keller, "Nature as 'Her.'"

10 Evelyn Fox Keller, "Pot-Holes Everywhere: How (Not) to Read my Biography of Barbara McClintock," in *Writing about Lives in Science: (Auto)Biography, Gender, and Genre*, ed. Paola Govoni and Zelda Alice Franceschi (Göttingen: V&R Unipress, 2014), 33–34; Keller, *Making Sense of My Life in Science*, 127.

11 Keller, "Pot-Holes Everywhere," 34.

12 Keller, "Pot-Holes Everywhere," 34.

13 Evelyn Fox Keller, *A Feeling for the Organism: The Life and Work of Barbara McClintock* (New York: Freeman, 1983), 16–17.

14 Keller, "Pot-Holes Everywhere," 34.

15 Keller, "Pot-Holes Everywhere," 34.

16 Keller, "Pot-Holes Everywhere," 35.

17 Keller, "Pot-Holes Everywhere," 35.

18 Keller, "Pot-Holes Everywhere," 36.

19 Keller, "Pot-Holes Everywhere," 36; Keller, *A Feeling for the Organism*, xii.

20 Keller, *A Feeling for the Organism*, 1.

21 Keller, *A Feeling for the Organism*, 29.

22 Keller, *A Feeling for the Organism*, 4

23 Keller, *A Feeling for the Organism*, 9.

24 Francis Crick first wrote of "the central dogma" in unpublished notes in 1956 and discussed it in a lecture in 1957. According to this idea, once information had flowed from DNA to proteins, it couldn't flow back in the other direction ("On Protein Synthesis," *Symposia of the Society for Experimental Biology* 12 (1958): 153.

25 Keller, *A Feeling for the Organism*, xii.

26 Keller, *A Feeling for the Organism*, xvii–xviii.

27 Keller, *A Feeling for the Organism*, 13.

28 Keller, "Pot-Holes Everywhere," 37.

29 Keller, "Pot-Holes Everywhere," 38.

30 Keller, "Pot-Holes Everywhere," 37. Thanks to Wendy Call for discussing this dynamic with me in September 2021.

31 Keller, *Reflections on Gender and Science*, 174.

32 Keller, "Pot-Holes Everywhere," 40.

33 Evelyn Fox Keller, "Feminism and Science," *Signs* 7, no. 3 (1982): 589–602. Also see Sandra Korn, "Sex, Science, and Politics in the Sociobiology Debate" (bachelor's thesis, Harvard University, March 2014).

34 Keller, "Feminism and Science," 589–602; Korn, "Sex, Science, and Politics in the Sociobiology Debate."

CHAPTER II

1 Kate Zernike, "The Reluctant Feminist," *New York Times*, April 8, 2001; Nancy Hopkins, "The High Price of Success in Science," *Radcliffe Quarterly* 10 (June 1976): 16–18.

2 Hopkins, "The High Price of Success in Science," 18.

3 Courtney Humphries, "Measuring Up," *MIT Technology Review*, August 16, 2017, https://www.technologyreview.com; Nancy Hopkins, "Mirages of Gender Equality (1960–2014)," speech at the fiftieth reunion of the Harvard-Radcliffe class of 1964, May 2014, http://hr1964.org.

4 Nancy Hopkins, "My Crush on DNA," *New York Times*, September 5, 2009; Christen Brownlee, "Biography of Nancy Hopkins," *Proceedings of the National Academy of Sciences (PNAS)* 101, no. 35 (2004): 12789; Hopkins, "Mirages of Gender Equality."

5 Hopkins, "Mirages of Gender Equality"; Christa Kuljian, interview with Nancy Hopkins, September 20, 2019; Neil A. Campbell, "Unit 3: Genetics," in *Biology*, 6th ed. (San Francisco: Cummings, 2005); Hopkins, "Mirages of Gender Equality." In my interview, I asked if Hopkins had approached Watson on the first day of class. She answered: "It's amazing that I don't remember when I first approached him to ask him if I could work in his lab. I remember a certain panic that the class would end and that I would be separated from this science. I couldn't live without it. I'm pretty sure it wasn't that [first] day. I wouldn't have had the courage. I don't think it would have occurred to me."

6 John Hockenberry interview with Nancy Hopkins, February 13, 2008, Infinite History Project, MIT, https://infinite.mit.edu.

7 Hopkins, "Mirages of Gender Equality"; Alicia Chen, "Women in the Sciences Still Struggle, Hopkins Says," *Brown Daily Herald*, October 22, 2009; Kuljian, interview with Hopkins.

8 Nancy Hopkins, "The Status of Women in Science and Engineering at MIT," presentation at MIT's 150th anniversary symposium, 2011, https://infinite.mit.edu.

9 James D. Watson, letter to Detlev Bronk, December 18, 1963, JDW/2/2/862/2, Cold Spring Harbor Laboratory Archives.

10 Kristin Kain, "Using Zebrafish to Understand the Genome: An Interview with Nancy Hopkins," *Disease Models and Mechanisms* 2 (May-June 2009): 214–17.

11 Kain, "Using Zebrafish to Understand the Genome."

12 Nancy Hopkins, "The High Price of Success in Science," *Radcliffe Quarterly* 10 (June 1976): 16–18.

13 John Hockenberry, interview with Nancy Hopkins, February 13, 2008, https://infinitehistory.mit.edu.

14 Nancy Hopkins, "The High Price of Success for Women in Science," 18.

15 Ruth Hubbard, review of *Rosalind Franklin and DNA*. *Signs* 2, no. 1 (1976): 230, 231, 234.

16 Kate Zernike, *The Exceptions: Nancy Hopkins, MIT, and the Fight for Women in Science* (New York: Scribner, 2023), 203–6.

17 Margaret Horton Weiler published several articles in the 1960s and contributed to *The Spectroscopy of Semiconductors* (1992). She is also mentioned in Donald Stevenson, Marion Reine, and Roshan L. Aggarwal, eds., *Benjamin Lax—Interviews on a Life in Physics at MIT* (New York: Routledge, 2020): "Margaret Horton Weiler (b. 1949) was hired by Ben in the fall of 1965 to work with Ken Button at the Magnet Lab to set up experiments with two lasers that she built for him (CO_2 and HCN). . . . She earned a PhD in physics in 1977 and then joined the MIT Physics Department faculty. . . . She would go on to have a distinguished and productive career in industrial physics, developing advanced HgCdTe infrared detectors for military and space applications."

18 Kuljian, interview with Hopkins; Hockenberry, interview with Nancy Hopkins.

19 Kuljian, interview with Hopkins; Hockenberry, interview with Nancy Hopkins.

20 Hopkins, "Mirages of Gender Equality."

21 Brownlee, "Biography of Nancy Hopkins," 12790.

22 Hopkins, "Mirages of Gender Equality."

CHAPTER 12

1 Clarence G. Williams, "Evelynn Hammonds," in *Technology and the Dream: Reflections on the Black Experience at MIT, 1941–1999* (Cambridge, MA: MIT Press, 2001), 939.

2 Bonnie Spanier, interview with Evelynn Hammonds, c. 1986, Bonnie Spanier Papers, Schlesinger Library, Harvard University.

3 Christa Kuljian, interview with Dolita Cathcart, March 9, 2019; Christa Kuljian, interview with Beverly Smith, November 27, 2020; Christa Kuljian, interview with Evelynn Hammonds, August 18, 2020. All three women recalled these regular evening meetings with young Black women in Cambridge.

4 Christa Kuljian, interview with Peggy McIntosh, October 17, 2022. McIntosh commented on how, ironically, the older white men who looked down on Hammonds had to rely on her for guidance in the new era of computer science. Also see Clarence Williams, "Evelynn Hammonds," 940.

5 Kuljian, interview with McIntosh. According to McIntosh, Hubbard may have suggested that Hammonds would be a great participant for the seminar. Also see Peggy McIntosh, letter to Anne Fausto-Sterling, April 18, 1984; and Anne Fausto-Sterling, letter to Peggy McIntosh, May 9, 1984; both in box 8, folder 8.4, Anne Fausto-Sterling Papers, Brown University.

6 Participants included Celia Alvarez, Margaret Andersen, Patricia Brown, Dorothy Buerk, Zala Chandler, Anne Fausto-Sterling, Evelynn Hammonds, Sandra Harding, Saj-nicole Joni, Evelyn Fox Keller, Barbara Kneubuhl, Judith Kroll, Helen Longino, Jan Roland Martin, Peggy McIntosh, Andree Nicola-McLaughlin, Katheryn Quina, Joan Rothschild, Patrocinio Schweickart, Gwyned Simpson, Bonnie Spanier, Trudy Ann Villars, Caroline Whitbeck, and Andrea Worthington (Bonnie Spanier, personal communication, April 6, 2021).

7 Nancy Hartsock, "The Feminist Standpoint: Developing the Call for Feminist Historical Materialism," in *Discovering Reality*, ed. Sandra Harding and Merryll B. Hintikka (New York: Springer, 1983), 283–310. The main idea of standpoint theory is that a person's perspective is shaped by their social and political context. Sandra Harding described strong objectivity as emphasizing the importance of bringing in the experiences and voices of those who have been marginalized in knowledge creation. Strong objectivity builds on standpoint theory ("Rethinking Standpoint Epistemology: What Is 'Strong Objectivity?,'" in *Feminist Theory: A Philosophical Anthology*, ed. Ann E. Cudd and Robin O. Andreasen [Oxford: Blackwell, 2005]).

8 Kuljian, interview with McIntosh.

9 Anne Fausto-Sterling, letter to Peggy McIntosh , May 9, 1984, box 8, folder 8.4, Anne Fausto-Sterling Papers, Brown University.

10 Christa Kuljian, interview with Anne Fausto-Sterling, January 26, 2021; Christa Kuljian, interview with Sandra Harding, April 19, 2021.

11 Spanier, interview with Hammonds; Christa Kuljian, interview with Evelynn Hammonds, November 19, 2020.

12 Spanier, interview with Hammonds.

13 Kuljian, interview with Fausto-Sterling.

14 Peggy McIntosh and Maggie Andersen, letters to Mellon Seminar participants, October 3, 1984, October 23, 1984, and May 3, 1985; Helen Longino, letter to Mellon Seminar participants, November 2, 1984; Shirley Malcom, letter to Mellon Seminar participants, September 26, 1984, all in box 8, folder 8.4, Anne Fausto-Sterling Papers, Brown University.

15 Peggy McIntosh and Maggie Andersen, letter to Mellon Seminar participants, February 7, 1985, box 8, folder 8.4; Anne Fausto-Sterling, notes on Evelynn Hammonds's presentation, titled "Science, Women of Color and Feminism," n.d.; both in Anne Fausto-Sterling Papers, Brown University. Hammonds had read Kenneth R. Manning, *Black Apollo of Science: The Life of Ernest Everett Just* (Oxford: Oxford University Press, 1983). See Sara P. Díaz, "Gender, Race, and Science: A *Feminista* Analysis of Women of Color in Science" (PhD diss., University of Washington, 2012); and Sara P. Díaz, "'A Racial Trust': Individualist, Eugenicist, and Capitalist Respectability in the Life of Roger Arliner Young," *Souls* 18, nos. 2–4 (2016): 235–62.

16 Anne Fausto-Sterling, letter to Peggy McIntosh, June 3, 1985, box 8, folder 8.4, Anne Fausto-Sterling Papers, Brown University.

17 Kuljian, interview with Harding.

18 Kuljian, interview with Harding.

19 Christa Kuljian, interview with Evelynn Hammonds, August 18, 2020; Evelynn Hammonds, "Women in STEM in the Twenty-First Century: The Stories We Tell and Action We Need," ARROWS Lecture, April 22, 2019, Boston University, https://www.bu.edu.

20 Kuljian, interview with Hammonds, August 18, 2020.

21 Kuljian, interview with Hammonds, November 19, 2021.

22 Hammonds's mother died on November 2, 1985, three days before her fifty-fifth birthday. The book based on Hammonds's dissertation is dedicated to her: *Childhood's Deadly Scourge* (Baltimore: Johns Hopkins University Press, 1999), viii–ix.

CHAPTER 13

1 J. C. [in-house staff writer], "Anne Fausto-Sterling," *Current Biography* (New York: Wilson, September 2005), 31.

2 Christa Kuljian, interview with Anne Fausto-Sterling, January 26, 2021.

3 Azeen Ghorayshi, "Conversations with Anne Fausto-Sterling," *Method*, November 12, 2015, www.methodquarterly.com.

4 Kuljian, interview with Fausto-Sterling.

5 Anne Fausto-Sterling, "The Myth of Neutrality: Race, Sex and Class in Science," *Radical Teacher* 19 (1981), 21.

6 Anne Fausto-Sterling and L. Hsieh, "Studies on the Female Sterile Mutant Rudimentary of Drosophila Melanogaster.1: An Analysis of the Rudimentary Wing Phenotype," *Developmental Biology* 51, no. 2 (1976): 269–81. I'm grateful to Nadja Cech for discussing this paper and its implications.

7 Anne Fausto-Sterling has stated that "her ambition is to restructure dichotomous conversations—inside the academy, in public discourse, and ultimately in the framing of social policy—in order to enable an understanding of the inseparability of nature/nurture" (https://www.annefaustosterling.com).

8 Fausto-Sterling, "The Myth of Neutrality," 22.

9 Fausto-Sterling, "The Myth of Neutrality," 22.

10 J. C., "Anne Fausto-Sterling," 31; Kuljian, interview with Fausto-Sterling.

11 Kuljian, interview with Fausto-Sterling.

12 Anne Fausto-Sterling and Evelynn Hammonds, "A Discussion about the Past and Future of Feminist Science Studies," May 2, 2014, https://www.annefaustosterling.com; Kuljian, interview with Fausto-Sterling.

13 Anne Fausto-Sterling, "Making Science Masculine: A Review of *The Mind Has No Sex?: Women in the Origins of Modern Science* by Londa Schiebinger," *Women's Review of Books* 7 (April 1990): 13–14; Anne Fausto-Sterling, "Race, Gender, and Science," *Transformations* 2 (Fall 1991): 7.

14 Kuljian, interview with Fausto-Sterling.

15 Margaret Rossiter, *Women Scientists in America*, vol. 1, *Struggles and Strategies to 1940* (Baltimore: Johns Hopkins University Press, 1982), xv.

16 Ruth Hubbard, review of *Women Scientists in America* by Margaret Rossiter, *Harvard Educational Review* 54, no. 4 (1984), 466.

17 Kuljian, interview with Fausto-Sterling.

18 Bonnie Spanier, interview with Evelynn Hammonds, c. 1986, Bonnie Spanier Papers, Schlesinger Library, Harvard University. At the time, Hammonds did not know about the African American women who had been working for NASA in Virginia. It would be another thirty years before Margot Lee Shetterly published *Hidden Figures* (New York: Morrow, 2016).

19 Kenneth Manning, *The Black Apollo of Science: The Life of Ernest Everett Just* (Oxford: Oxford University Press, 1985); Christa Kuljian, interview with Evelynn Hammonds, April 13, 2021.

20 Spanier, interview with Hammonds.

21 Another important book that came out that year was Willie Pearson Jr., *Black Scientists, White Society, and Colorless Science: A Study of Universalism in Science* (Millwood, NY: Associated Faculty Press, 1985).

22 Anne Fausto-Sterling and Lydia L. English, "Women and Minorities in Science: An Interdisciplinary Course," *Radical Teacher* 30 (January 1986): 19.

23 Kuljian, interview with Hammonds, April 13, 2021.

24 Spanier, interview with Hammonds.

25 Anne Fausto-Sterling, *Myths of Gender* (New York City: Basic Books, 1985), 4, 8.

26 Fausto-Sterling, *Myths of Gender*, 53–54.

27 Fausto-Sterling, *Myths of Gender*, 85.

28 Fausto-Sterling, *Myths of Gender*, 205–222.

29 Spanier, interview with Hammonds.

30 For a full description of the course and its syllabus, see Nancy Goddard and Mary Sue Henifin, "A Feminist Approach to the Biology of Women," *Women's Studies Quarterly* 12, no. 4 (1984): 11–18. Also see Bonnie Spanier, "Women's Studies and the Natural Sciences: A Decade of Change," *Frontiers* 8, no. 3 (1986), 66–72. Spanier mentions Hampshire College's Women in Science Group and the conferences that started in 1983 (70).

31 E. Frances White, curriculum vita, https://nyu.academia.edu; Darlene Clark Hine and Elsa Barkley Brown, *Black Women in America: An Historic Encyclopedia* (New York: Carlson, 1993); Kenneth R. Manning, "Roger Arliner Young, Scientist" *Sage*, 6 (1989): 3–7; Sara P. Díaz, "'A racial trust': Individualist, Eugenicist, and Capitalist Respectability in the Life of Roger Arliner Young," *Souls*, October 1, 2016, 235–62.

32 Evelynn Hammonds, letter to Evelyn Fox Keller, February 2, 1987, box 6, folder 3, Evelyn Fox Keller Papers, Schlesinger Library, Harvard University.

33 Hammonds, letter to Keller.

34 Hammonds, letter to Keller; E. Frances White and Ann Woodhull-McNeal, "Challenging the Scientific Myths of Gender and Race: A Review of *Myths of Gender,*" *Radical America* 20, no. 4 (1986): 25–32.

35 White and Woodhull-McNeal, "Challenging the Scientific Myths of Gender and Race," 25–32.

36 Evelynn Hammonds, "A Radio Profile of Audre Lorde," *Jennifer Abod Radio Show* (Boston: WGBH, 1988), http://www.jenniferabod.com.

37 Hammonds, "A Radio Profile of Audre Lorde."

38 Evelynn Hammonds, letter to Carl Kaysen, January 6, 1987, box 6, folder 3, Evelyn Fox Keller Papers, Schlesinger Library, Harvard University.

39 Kuljian, interview with Hammonds, April 13, 2021; Hammonds, letter to Keller.

40 Aimee Sands, "Never Meant to Survive: A Black Woman's Journey: An Interview with Evelynn Hammonds," in *The Racial Economy of Science: Toward a Democratic Future*, ed. Sandra Harding (Bloomington: Indiana University Press, 1993), 239–24.

41 Audre Lorde, "A Litany for Survival," in *The Black Unicorn: Poems* (New York: Norton, 1978), 31–32.

42 Christa Kuljian, interview with Evelynn Hammonds, August 18, 2020.

43 Kuljian, interview with Hammonds, August 18, 2020.

44 Londa Schiebinger, "Introduction: Feminism inside the Sciences," *Signs* 28 (Spring 2003): 859; Banu Subramaniam, "Moored Metamorphoses: A Retrospective Essay on Feminist Science Studies," *Signs* 34 (Summer 2009): 951.

45 Rita Arditti, "Book Review: *The Death of Nature* by Carolyn Merchant," *Science for the People* 15 (January–February 1983): 44–45.

46 Rita Arditti, "Feminism and Science," in *Science and Liberation*, ed. Rita Arditti, Pat Brennan, and Steve Cavrak (Boston, South End Press, 1980), 367.

47 Evelynn Hammonds, "Black (W)holes and the Geometry of Black Female Sexuality," *differences* 6, nos. 2–3 (1994): 143.

48 Colleen Walsh, "Stonewall Then and Now," *Harvard Gazette*, June 27, 2019. Also, Hammonds shared that she had come out as a lesbian by the time she'd started her PhD in *Secret Court 100: Black Queer Harvard*, August 20, 2020, https://www.youtube.com.

49 Kuljian, interview with Hammonds, August 18, 2020

50 Kuljian, interview with Hammonds, August 18, 2020.

51 Ruth Hubbard, letter to Margaret Randall, July 26, 1987, in Ruth Hubbard and Margaret Randall, *The Shape of Red: Insider/Outsider Reflections* (San Francisco: Cleis, 1988), 20–21.

52 Kuljian, interview with Hammonds, April 13, 2021; Kuljian, interview with Fausto-Sterling.

53 Evelynn Hammonds, "Women in STEM in the Twenty-First Century: The Stories We Tell and Action We Need," ARROWS Lecture, April 22, 2019, Boston University, https://www.bu.edu.

54 Kuljian, interview with Hammonds, April 13, 2021.

55 Kuljian, interview with Hammonds, April 13, 2021.

56 Evelynn Hammonds, "Race, Sex, AIDS: The Construction of 'Other,'" *Radical America* 20, no. 6 (1987): 28–36.

57 Evelynn Hammonds, presentation at "Radical Roots Symposium," MIT, March 24, 2021; Kuljian, interview with Hammonds, April 13, 2021.

58 After completing her corrections, Hammonds earned her PhD in 1993. She mentions the graduate course she taught in Evelynn Hammonds and Banu Subramaniam, "A Conversation on Feminist Science Studies," *Signs* 28 (Spring 2003): 924.

CHAPTER 14

1 Constantine B. Simonides, letter to Evelyn Fox Keller, April 9, 1987, box 6, folder 4; Ford Ebner, letter to Evelyn Fox Keller, December 16, 1985, box 4, folder 5; Evelyn Fox Keller, letter to Ford Ebner, January 20, 1986, box 4, folder 5; Evelynn Hammonds, letter to Evelyn Fox Keller, January 6, 1987, box 6, folder 3; all in Evelyn Fox Keller Papers, Schlesinger Library, Harvard University

2 Ruth Hubbard, letters to Evelyn Fox Keller, September 9, 1987, March 30, 1988, April 20, 1988, August 12, 1988; Evelyn Fox Keller, letter to Ruth Hubbard, September 15, 1987; all in box 1, folder 22, Evelyn Fox Keller Papers, Schlesinger Library, Harvard University.

3 Ruth Hubbard, letters to Evelyn Fox Keller, September 9, 1987, March 30, 1988, April 20, 1988, and August 12, 1988; Evelyn Fox Keller, letter to Ruth Hubbard, September 15, 1987; all in box 1, folder 22, Evelyn Fox Keller Papers, Schlesinger Library, Harvard University.

4 Christa Kuljian, interview with Evelyn Fox Keller, September 17, 2019.

5 Hubbard, letters to Keller, September 9, 1987, March 30, 1988, April 20, 1988, and August 12, 1988.

6 Christa Kuljian, interview with Sandra Harding, April 19, 2021.

7 Margaret Burnham, comments at Ruth Hubbard's retirement symposium, June 2, 1990, Vt-232_0001, Ruth Hubbard Papers, Schlesinger Library, Harvard University.

8 Richard C. Lewontin, comments at Hubbard's retirement symposium.

9 Richard C. Lewontin, "Women versus the Biologists," *New York Review of Books*, April 7, 1994.

10 Ayofemi Folayan and Joanne Stato, "I Am Your Sister: A Tale of Two Conferences," *off our backs* 20, no. 11 (1990): 1–5, 9–11; fundraising letter, May 12, 1990, https://go.gale.com.

11 Helen Longino and Evelynn Hammonds, "Conflicts and Tensions in the Feminist Study of Gender and Science," in *Conflicts in Feminism*, ed. Marianne Hirsch and Evelyn Fox Keller (New York: Routledge, 1990), 164.

12 Longino and Hammonds, "Conflicts and Tensions," 176.

13 Longino and Hammonds, "Conflicts and Tensions," 177.

14 Longino and Hammonds, "Conflicts and Tensions," 178.

15 Longino and Hammonds, "Conflicts and Tensions," 178–79.

16 Longino and Hammonds, "Conflicts and Tensions," 179.

17 Longino and Hammonds, "Conflicts and Tensions," 180.

18 Longino and Hammonds, "Conflicts and Tensions," 180–81.

19 Evelyn Fox Keller, "Conclusion," in *Conflicts in Feminism*, 380–85.

20 Donna Haraway, interviewed by Banu Subramaniam, for "Foundations and Futures of Feminist Technoscience," *Catalyst*'s fifth anniversary webinar, April 22, 2021, https://catalystjournal.org.

21 Evelynn Hammonds and Banu Subramaniam, "A Conversation on Feminist Science

Studies," *Signs* 28 (Spring 2003): 923–44, 925; Bonnie Spanier, "Women's Studies and the Natural Sciences: A Decade of Change," *Frontiers* 8, no. 3 (1986): 66–72.

CHAPTER 15

1 Banu Subramaniam, "A Contradiction in Terms," *Women's Review of Books* 25 (February 1998): 25–26, 25.

2 Subramaniam, "A Contradiction in Terms," 25.

3 Subramaniam, "A Contradiction in Terms," 25.

4 Subramaniam, "A Contradiction in Terms," 25.

5 Christa Kuljian, interview with Banu Subramaniam, August 13, 2020; Ruth Hubbard, "Have Only Men Evolved?," in *Women Look at Biology Looking at Women* (Cambridge, MA: Schenkman, 1979); Sharon Traweek, *Beamtimes and Lifetimes* (Cambridge, MA: Harvard University Press, 1988); Banu Subramaniam, personal communication, June 1, 2023. The two graduate students were Jim Bever and Peggy Schultz.

6 Banu Subramaniam, *Ghost Stories for Darwin: The Science of Variation and the Politics of Diversity* (Urbana: University of Illinois Press, 2014), xiii–ix.

7 Catherine Russo, video interview with Rita Arditti, c. 1991, http://ritaarditti.com.

8 Russo, video interview with Arditti.

9 Emily Martin, "The Egg and the Sperm: How Science Has Constructed a Romance Based on Stereotypical Male-Female Roles," *Signs* 16 (Spring 1991): 486.

10 Emily Martin, personal communication, January 19, 2021, and January 20, 2021.

11 Vernon B. Mountcastle, *Medical Physiology*, 14th ed. (London: Mosby, 1980), 2: 1624, quoted in Martin, "The Egg and the Sperm," 486; italics are Martin's.

12 Emily Martin, "The Egg and the Sperm," 487.

13 Martin, "The Egg and the Sperm," 489–90.

14 Martin, "The Egg and the Sperm," 492.

15 Martin, "The Egg and the Sperm," 496.

16 Martin, "The Egg and the Sperm," 501.

17 Martin, personal communication, January 20, 2021; Christa Kuljian, interview with Anne Fausto-Sterling, January 26, 2021.

18 Emily Martin, *The Woman in the Body: A Cultural Analysis of Reproduction* (Boston: Beacon, 1992), xxi.

19 Martin, *The Woman in the Body*, xxii.

20 Biology and Gender Study Group [Athena Beldecos, Sarah P. Baily, Scott F. Gilbert, Karen A. Hicks, Lori J. Kenschaft, Nancy A. Niemczyk, Stephanie A. Schaertel, and Andrew B. Wedel], "The Importance of Feminist Critique for Contemporary Cell Biology," *Hypatia* 3 (Spring 1988): 61–76.

21 Biology and Gender Study Group, "The Importance of Feminist Critique," 68–69.

22 Kenneth R. Manning, *The Black Apollo of Science: The Life of Ernest Everett Just* (Oxford: Oxford University Press, 1983), 263; Biology and Gender Study Group, "The Importance of Feminist Critique," 69.

23 Biology and Gender Study Group, "The Importance of Feminist Critique," 68–69.

24 Biology and Gender Study Group, "The Importance of Feminist Critique," 71.

25 As of 2023, the book was in its thirteenth edition.

26 Christa Kuljian, interview with Emily Martin, January 20, 2021.

CHAPTER 16

1 Simon LeVay, "A Difference in Hypothalamic Structure between Homosexual and Heterosexual Men," *Science*, August 30, 1991, 1034–37. This article engendered a great deal of commentary, both scientific and political, during the next decade.

2 Anne Fausto-Sterling, *Myths of Gender*, 2nd ed. (New York City: Basic Books, 1992), vi–vii.

3 Fausto-Sterling, *Myths of Gender*, 256. Fausto-Sterling concluded that the methodology and data in LeVay's study were flawed. She also concluded that homosexuals should have full legal and social rights and argued that LeVay's preliminary findings might be used with malevolent intent: "Both biological and environmental views about the origin of homosexuality can be used to oppress gay people. Only the continued fight to increase acceptance of homosexuality, regardless of its origins, can help in the end. . . . Human sexual behaviors have complex origins. It seems unlikely that there is a single cause of homosexuality, rather it is a behavior endpoint that can be reached by many different paths. . . . Only by leaving behind fixed, linear models of the brain and behavior and progressing to complex, plastic networked approaches will we get somewhere" (ibid.). LeVay wrote that "sexual orientation in humans is amenable to study at the biological level, and this discovery opens the door to the study of neurotransmitters or receptors that might be involved in regulating this aspect of personality. Further interpretation of the results of this study must be considered speculative" ("A Difference in Hypothalamic Structure," 1036).

4 Anne Fausto-Sterling, "Science Matters, Culture Matters," *Perspectives in Biology and Medicine* 46, no. 1 (2003): 109–24.

5 Fausto-Sterling, "Science Matters, Culture Matters," 110–12.

6 Fausto-Sterling, "Science Matters, Culture Matters," 112.

7 Fausto-Sterling, "Science Matters, Culture Matters," 115.

8 Fausto-Sterling, "Science Matters, Culture Matters," 122.

9 J. C. [in-house staff writer], "Anne Fausto-Sterling," *Current Biography* (New York: Wilson, September 2005), 35.

10 Anne Fausto-Sterling, *Sexing the Body*, 2nd ed. (New York: Basic Books, 2020), ix.

11 Anne Fausto-Sterling, "The Five Sexes: Why Male and Female Are Not Enough," *The Sciences* 33 (March–April 1993): 20–24; Anne Fausto-Sterling, "The Five Sexes Revisited," *The Sciences* 40 (July–August 2000): 19.

12 Subramaniam, "A Contradiction in Terms," 25.

13 Anne Fausto-Sterling, "Building Two-Way Streets: The Case of Feminism and Science," *National Women's Studies Association Journal* 4, no. 3 (1992): 336.

14 Anne Fausto-Sterling, "Building Two-Way Streets," 337.

15 Subramaniam, "A Contradiction in Terms," 26; Christa Kuljian, interview with Banu Subramaniam, August 13, 2020.

16 Subramaniam, "A Contradiction in Terms," 26.

17 Bonnie Spanier, interview with Evelynn Hammonds, c. 1986, Bonnie Spanier Papers, Schlesinger Library, Harvard University.

18 Fausto-Sterling, "Building Two-Way Streets," 339.

19 James D. Watson, *The Double Helix* (New York: Norton, 1980), 15.

20 Bonnie Spanier, *Im/partial Science: Gender Ideology and Molecular Biology* (Bloomington: Indiana University Press, 1995).

21 Ruth Hubbard and Elijah Wald, *Exploding the Gene Myth* (Boston: Beacon, 1993);

Evelyn Fox Keller, *Reconfiguring Life: Metaphors of Twentieth-Century Biology* (New York: Columbia University Press, 1995); Evelyn Fox Keller, *The Century of the Gene* (Cambridge, MA: Harvard University Press, 2000).

22 Ruth Hubbard, "Comments on Anne Fausto-Sterling's 'Building Two-Way Streets,'" *National Women's Studies Association Journal* 5 (Spring 1993): 47.

23 Sandra Harding, "Comments on Anne Fausto-Sterling's 'Building Two-Way Streets,'" *National Women's Studies Association Journal* 5 (Spring 1993): 52.

24 Anne Fausto-Sterling, "Comments on Anne Fausto-Sterling's 'Building Two-Way Streets,'" *National Women's Studies Association Journal* 5 (Spring 1993): 79.

CHAPTER 17

1 Evelynn Hammonds, *Secret Court 100: Black Queer Harvard*, August 20, 2020, https://www.youtube.com.

2 Clarence G. Williams, "Evelynn Hammonds," in *Technology and the Dream: Reflections on the Black Experience at MIT, 1941–1999* (Cambridge: MIT Press, 2001), 941.

3 For more on Roger Arliner Young, see Kenneth R. Manning, "Roger Arliner Young, Scientists" *Sage* 6, (1989): 3-7; Sara P. Díaz, "Gender, Race, and Science: A *Feminista* Analysis of Women of Color in Science" (PhD diss., University of Washington, 2012); and Sara P. Díaz, "'A racial trust': Individualist, Eugenicist, and Capitalist Respectability in the Life of Roger Arliner Young," *Souls* 18, nos. 2–4 (2016): 235–62.

4 Christa Kuljian, interview with Evelynn Hammonds, November 19, 2021; Evelynn Hammonds, , "Race, Gender, and the History of Women in Science," paper presented at the History of Science Society meeting, November 11, 1993, box 42, Evelynn Hammonds Papers, Harvard University Archives.

5 In a letter to Evelyn Fox Keller, Evelynn Hammonds wrote: "Anne Fausto-Sterling gave me an interesting lead. . . . I want to try to weave in an analysis of the book with the data in the paper about stereotypes of Black women" (April 29, 1991, Evelyn Fox Keller Papers, Schlesinger Library, Harvard University). Also see Marjorie Hill Alle, *The Great Tradition* (Boston: Houghton Mifflin, 1937), mentioned in Hammonds, "Race, Gender, and the History of Women in Science."

6 Kuljian, interview with Hammonds, November 19, 2021. On the first published books about Black women in science, see Wini Warren, *Black Women Scientist in the United States* (Bloomington: Indiana University Press, 1999); and Diann Jordan, *Sisters in Science: Conversations with Black Women Scientists on Race, Gender, and Their Passion for Science* (West Lafayette, IN: Purdue University Press, 2006).

7 Williams, "Evelynn Hammonds"; Kuljian, interview with Hammonds, November 19, 2021.

8 Scholars contributing to these fields include Dorothy Roberts, Alondra Nelson, Harriet Washington, Keith Waimoo, Ruha Benjamin, Linda Villarosa, Deirdre Cooper Owen, Rana Hogarth, Safiya Noble, and Joy Buolamwini.

9 Evelynn Hammonds, "Who Speaks for Black Women?," *Sojourner* 17 (November 1991): 7. Today, Hammonds might use Moya Bailey's term *mysogynoir* (*Misogynoir Transformed: Black Women's Digital Resistance* [New York: New York University Press, 2021]).

10 Evelynn Hammonds, letter to Audre Lorde, December 5, 1991, box 2, folder 50, Audre Lorde Papers, Women's Research and Resource Center, Spelman College.

11 Hammonds, "Who Speaks for Black Women?," 8; "African American Women in Defense of Ourselves," *New York Times*, November 17, 1991.

12 Evelynn Hammonds, letters to Audre Lorde, December 5, 1991, and November 17, 1992, both in box 2, folder 50, Audre Lorde Papers, Women's Research and Resource Center, Spelman College.

13 Saidiya Hartman, "The Territory between Us: A Report on 'Black Women in the Academy: Defending Our Name: 1894–1994,'" *Callaloo* 17 (Spring 1994): 439–49.

14 Lorraine O'Grady, "Olympia's Maid: Reclaiming Black Female Subjectivity," *Afterimage* 20, no, 1 (1992): 14.

15 Evelynn Hammonds, "Black (W)holes and the Geometry of Black Female Sexuality," *differences* 6 (Summer–Fall 1994): 126. Also, see forthcoming 2024 issue of *differences* celebrating the 30th anniversary of this essay.

16 Christa Kuljian, interviews with Evelynn Hammonds, August 18, 2020, and April 13, 2021.

17 Kuljian, interview with Hammonds, April 13, 2021; Jenny Reardon, *Race to the Finish: Identity and Governance in the Age of Genomics* (Princeton, NJ: Princeton University Press, 2004); Lundy Braun, *Breathing Race into the Machine: The Surprising Career of the Spirometer from Plantation to Genetics* (Minneapolis: University of Minnesota Press, 2014); Alondra Nelson, *Body and Soul: The Black Panther Party and the Fight against Medical Discrimination* (Minneapolis: University of Minnesota Press, 2013). Also see Alondra Nelson, *The Social Life of DNA: Race Reparations and Reconciliation after the Genome* (Boston: Beacon, 2016)

18 Anne Fausto-Sterling, "Gender, Race, and Nation: The Comparative Anatomy of 'Hottentot' Women in Europe, 1815–1817," in *Deviant Bodies: Critical Perspectives on Difference in Science and Popular Culture*, ed. Jennifer Terry and Jacqueline Erla (Bloomington: Indiana University Press, 1995).

19 Christa Kuljian, interview with Banu Subramaniam, August 13, 2020.

20 Donna Haraway, *Primate Visions: Gender, Race, and Nature in the World of Modern Science* (New York: Routledge, 1989); Nancy Stepan, "Race and Gender: The Role of Analogy in Science," in *Anatomy of Racism*, ed. David Goldberg (Minneapolis: University of Minnesota Press, 1990); Sandra Harding, ed., *The "Racial" Economy of Science* (Bloomington: Indiana University Press, 1993).

21 Patricia Hill Collins, "Moving beyond Gender: Intersectionality and Scientific Knowledge," in *Revisioning Gender*, ed. Myra Marx Ferree, Judith Lorber, and Beth B. Hess. (Walnut Creek, CA: Altamira, 2000), 263.

22 Collins, "Moving beyond Gender." 268.

23 Evelynn Hammonds and Banu Subramaniam, "A Conversation on Feminist Science Studies," *Signs* 28 (Spring 2003): 928.

24 Hammonds and Subramaniam, "A Conversation on Feminist Science Studies," 928–29.

25 Hammonds and Subramaniam, "A Conversation on Feminist Studies," 928.

26 Robert Merton, "The Matthew Effect in Science," *Science*, January 5, 1968, 56–63.

27 Margaret Rossiter, "The Matthew Matilda Effect in Science," *Social Studies of Science* 23, no. 2 (1993): 325–41.

28 Rossiter, "The Matthew Matilda Effect in Science," 330.

29 Patricia Farnes, "Women in Medical Science," in *Women of Science: Righting the Record*, ed. Gabriele Kass-Simon and Patricia Farnes (Bloomington: Indiana University Press, 1993), 289.

CHAPTER 18

1 Christa Kuljian, interview with Sandra Harding, April 19, 2021; Michael Harris, "Science Wars: The Next Generation," *Science for the People* 22, no. 1 (2021), https://magazine.science forthepeople.org.

2 Paul Gross and Norman Levitt, *Higher Superstition: The Academic Left and Its Quarrels with Science*, 2nd ed. (Baltimore: Johns Hopkins University Press, 1998), 4.

3 Ruth Hubbard, "Gender and Genitals: Constructs of Sex and Gender, *Social Text* 46/47 (Spring–Summer 1996): 157–65.

4 Ruth Hubbard, "The Tunnel," journal entry, May–September 1997, box 17, folder 25, Ruth Hubbard Papers, Schlesinger Library, Harvard University.

5 Ruth Hubbard, "Facts and Feminism," *Science for the People* 18, no. 2 (1986): 16–20, 26. A later version of this article was published as "Science, Facts, and Feminism," *Hypatia* 3, no. 1 (1988): 5–17.

6 Richard C. Lewontin, "Women versus the Biologists," *New York Review of Books*, April 7, 1994.

7 Kuljian, interview with Harding.

8 Kuljian, interview with Harding.

9 Harris, "Science Wars."

10 Evelyn Fox Keller, *Making Sense of My Life in Science: A Memoir* (Amherst, MA: Modern Memoirs, 2023), 171–72.

11 Donna Haraway, "Manifesto for Cyborgs: Science, Technology, and Socialist Feminism in the 1980s," *Socialist Review* 80 (1985): 65–108; Donna Haraway, "Situated Knowledges: The Science Question in Feminism and the Privilege of Partial Perspective," *Feminist Studies* 14 (Autumn 1988): 575–99; Donna Haraway, *Primate Visions: Gender, Race, and Nature in the World of Modern Science* (New York: Routledge, 1989).

12 Sadly, the written text of Donna Haraway's talk was lost, but many attendees remember it clearly. See Yesmar Oyarzun and Aadita Chaudhury, "Interview with Donna Haraway on 4S Conference Blog," August 27, 2019, https://stsinfrastructures.org.

13 Oyarzun and Chaudhury, "Interview with Donna Haraway on 4S Conference Blog."

14 Ann Fausto-Sterling spoke about Donna Haraway's talk in a conversation with Evelyn Hammonds (symposium to honor Anne Fausto-Sterling, Brown University, May 2, 2014, https://www.annefaustosterling.com). Also Christa Kuljian, interview with Anne Fausto-Sterling, January 26, 2021; Christa Kuljian, interview with Evelynn Hammonds, November 19, 2021.

15 Keller, *Making Sense of My Life in Science*, 174.

16 Evelyn Fox Keller, *The Century of the Gene* (Cambridge, MA: Harvard University Press, 2000), 54–72.

CHAPTER 19

1 Courtney Humphries, "Measuring Up," *MIT Technology Review*, August 16, 2017, https://www.technologyreview.com.

2 Bonnie Spanier, interview with Nancy Hopkins, April 12, 1988, box 2, folder 4, Bonnie Spanier Papers, Schlesinger Library, Harvard University.

3 Christa Kuljian, interview with Nancy Hopkins, September 20, 2019; Christa Kuljian, interview with Bonnie Spanier, March 30, 2021; Spanier, interview with Hopkins.

4 Kuljian, interview with Spanier.

5 Nancy Hopkins, personal communication, July 31, 2023.

6 Kuljian, interview with Hopkins.

7 Kristen Kain, "Using Zebrafish to Understand the Genome: An Interview with Nancy Hopkins," *Disease Models and Mechanisms* 2 (May–June 2009): 214–17; Christen Brownlee, "Biography of Nancy Hopkins," *Proceedings of the National Academy of Science*, August 31, 2004, 12789–91.

8 Kuljian, interview with Hopkins.

9 Brownlee, "Biography of Nancy Hopkins," 12790; "Interview with Nancy Hopkins," *Campbell Essential Biology*, 6th ed., by Eric J. Simon, Jean L. Dickey, Jane B. Reece, and Kelly A. Hogan (Indianapolis: Pearson, 2005), unit 3. Nusslein-Volhard's lab work identified the genes needed for the development of a fly embryo. Her researchers treated flies with a mutagen—that is, something that can cause a mutation in DNA. Then they looked to see how the descendants of the treated flies developed. They damaged one gene at a time, or just a few genes, to see what happened when those genes were no longer functional. That way, they could identify which genes were necessary for the development of the fly embryo.

10 Kain, "Using Zebrafish to Understand the Genome."

11 John Hockenberry, interview with Nancy Hopkins, MIT Infinite History Project, February 13, 2008, https://infinitehistory.mit.edu; Brownlee, "Biography of Nancy Hopkins"; Humphries, "Measuring Up," 2017.

12 Hockenberry, interview with Hopkins.

13 Nancy Hopkins, personal communication, July 31, 2023.

14 Nancy Hopkins, "Mirages of Gender Equality (1960–2014)," speech at the fiftieth reunion of the Harvard-Radcliffe class of 1964, May 2014, http://hr1964.org.

15 Hopkins, "Mirages of Gender Equality."

16 Hopkins, "Mirages of Gender Equality."

17 Nancy Hopkins "Women of MIT: The Status of Women in Science and Engineering at MIT," speech, March 28, 2011, https://infinite.mit.edu.

18 Hopkins, "Mirages of Gender Equality." At Harvard, 5 percent of the faculty were women.

19 Hopkins, "Mirages of Gender Equality," 3. One woman had won the U.S. National Medal of Science by 1999; the other three won after 1999.

20 Hopkins, "Mirages of Gender Equality"; Hockenberry, interview with Hopkins; Kate Zernike, *The Exceptions: Nancy Hopkins, MIT, and the Fight for Women in Science* (New York: Scribner, 2023), 329–31.

21 Hockenberry, interview with Hopkins.

22 Hopkins, "Women of MIT."

23 Hockenberry, interview with Hopkins.

24 Hockenberry, interview with Hopkins.

25 Helen Longino and Evelynn Hammonds, "Conflicts and Tensions in the Feminist Study of Gender and Science," *Conflicts in Feminism*, ed. Marianne Hirsch and Evelyn Fox Keller (New York: Routledge, 1990), 164–83.

26 Hockenberry, interview with Hopkins.

27 Hockenberry, interview with Hopkins; Hopkins, "Mirages of Gender Equality"; Charles M. Vest, "Introductory Comments," in "A Study on the Status of Women Faculty in Science at MIT," *MIT Faculty Newsletter* 11 (March 1999), https://web.mit.edu.

28 Kuljian, interview with Hopkins.

29 Hockenberry, interview with Hopkins.

30 Evelynn Hammonds, "Women in STEM in the Twenty-First Century: The Stories We Tell and Action We Need," ARROWS lecture, Boston University, April 22, 2019, https://www.bu.edu.

31 Nancy Hopkins, "Experience of Women at the Massachusetts Institute of Technology," address to the Chemical Sciences Roundtable, National Research Council, Washington, D.C., 2000, https://www.ncbi.nlm.nih.gov.

32 Hopkins, "Experience of Women at the Massachusetts Institute of Technology."

33 Christa Kuljian, interview with Evelynn Hammonds, August 18, 2020.

34 Evelynn Hammonds, "Women in STEM in the Twenty-First Century."

35 Hopkins, personal communication.

36 Christa Kuljian, interview with Anne Fausto-Sterling, January 26, 2021.

37 Sarah H. Wright, "Minority Women Faculty Conferees Discuss Professional Obstacles," *MIT News*, January 30, 2002.

38 Hopkins, "Mirages of Gender Equality."

CHAPTER 20

1 "Leaders of 9 Universities and 25 Women Faculty Meet at MIT, Agree to Equity Reviews," *MIT News*, January 30, 2001.

2 John Hockenberry, interview with Nancy Hopkins, MIT Infinite History Project, February 13, 2008, https://infinitehistory.mit.edu.

3 Christa Kuljian, interview with Evelynn Hammonds, November 19, 2021.

4 Kuljian, interview with Hammonds, November 19, 2021.

5 Kuljian, interview with Hammonds, November 19, 2021.

6 "Evelynn M. Hammonds," *History Makers: The Digital Repository for the Black Experience*, 2005, https://www.thehistorymakers.org.

7 J. C. [in-house staff writer], "Anne Fausto-Sterling," in *Current Biography* (New York: Wilson, September 2005), 35.

8 Jonathan Finer, Alan Cooperman, and Fred Barbash, "America's First Gay Marriage," *Washington Post*, May 17, 2004.

9 Colleen Walsh, "Stonewall Then and Now," *Harvard Gazette*, June 27, 2019; Joanna Weiss and Anand Vaishnav, "Celebrations Envelop Cambridge City Hall," *Boston Globe*, May 17, 2004.

10 Hockenberry, interview with Hopkins.

11 Hockenberry, interview with Hopkins.

12 Hockenberry, interview with Hopkins.

13 Lawrence Summers, "Full Transcript: President Summers' Remarks at the National Bureau of Economic Research, January 14, 2005," *Harvard Crimson*, February 18, 2005.

14 Summers, "Full Transcript."

15 Summers, "Full Transcript."

16 Hockenberry, interview with Hopkins; Kate Zernike also describes this scene in *The Exceptions: Nancy Hopkins, MIT, and the Fight for Women in Science* (New York: Scribner, 2023), 350.

17 Summers, "Full Transcript."

18 Steven Pinker, *The Blank Slate: The Modern Denial of Human Nature* (New York: Viking, 2002), 337–71.

19 Pinker, *The Blank Slate*, 341, 352, 354.

20 Marcella Bombardieri, "Summers' Remarks on Women Draw Fire," *Boston Globe*, January 17, 2005.

21 George F. Will, "Opinion: Harvard Hysterics," *Washington Post*, January 26, 2005.

22 Stephen Marks, "Low Female Tenure Numbers Decried in Letter to Summers," *Harvard Crimson*, September 23, 2004.

23 Robert Hayden, interview with Evelynn Hammonds, January 29, 2005, *The History Makers: The Digital Repository for the Black Experience*, https://www.thehistorymakers.org.

24 Lawson, Joanne, "Harvard President Condemned for 'Offensive' Speech" *The Guardian*, January 18, 2005.

25 Bombardieri, "Harvard Women's Group Rips Summers," *Boston Globe*, January 19, 2005.

26 Hana Rachel Alberts, "On Lawrence Summers, Women, and Science: Changing Debates about the Biology of Sex Differences at Harvard Since 1969" (honors degree thesis, Harvard University, 2006).

27 Alberts, "On Lawrence Summers, Women, and Science."

28 Alberts, "On Lawrence Summers, Women, and Science," 77, 79–80; Hayden, interview with Hammonds.

29 Alberts, "On Lawrence Summers, Women, and Science," 41.

30 Mazarin Banaji, Elizabeth Spelke, and Nancy Hopkins, "Impediments to Change: Revisiting the Women in Science Question," presentation at the Radcliffe Institute for Advanced Study, March 21, 2005, audiotape, Schlesinger Library, Harvard University; Alberts, "On Lawrence Summers, Women, and Science," 62–63.

31 Evelyn Fox Keller, "Innate Confusions: Nature, Nurture, and All That," presentation at the Radcliffe Institute for Advanced Study, April 7, 2005, Schlesinger Library, Harvard University; Alberts, "On Lawrence Summers, Women, and Science," 64. Also see Evelyn Fox Keller, *The Mirage of a Space between Nature and Nurture* (Durham: Duke University Press, 2010).

32 Many scientists have written about this interaction between genes and their environment, including Ruth Hubbard and Elijah Wald, *Exploding the Gene Myth* (Boston: Beacon, 1993); and Evelyn Fox Keller, *The Century of the Gene* (Cambridge, MA: Harvard University Press, 2000).

33 Stephen S. Hall, "Revolution Postponed: Why the Human Genome Project has been Disappointing," *Scientific American*, October 1, 2010; Alasdair Mackenzie and Andreas Kolb, "The Human Genome at 20: How Biology's Most Hyped Breakthrough Led to an Anti-Climax and Arrests," *The Conversation*, February 19, 2021.

34 Alberts, "On Lawrence Summers, Women, and Science," 87.

35 Christa Kuljian, interview with Anne Fausto-Sterling, January 26, 2021; Christa Kuljian, interview with Banu Subramaniam, September 30, 2019.

36 Alan Finder, Patrick D. Healy, and Kate Zernike, "President of Harvard Resigns, Ending Stormy 5-Year Tenure," *New York Times*, February 22, 2006. After Summers left Harvard, he worked at a hedge fund and then at a few major banks before taking various economic-policy positions in the Obama administration. In September 2004, Summers and Pinker, along with several other Harvard colleagues, including Alan Dershowitz, met with Jeffrey Epstein (later publicly exposed as a serial sexual predator), who had donated $6 million to found the Institute for Theoretical Biology at Harvard. Martin Nowak, the head of the new institute, invited Pinker and Robert Trivers, a proponent of sociobiology and a former colleague of Wilson's, to have lunch with Epstein, who was interested in their work. A photographer, Rick Friedman, attended to take professional photographs.

37 Evelynn Hammonds, "Women in STEM in the Twenty-First Century: The Stories We Tell and Action We Need," ARROWS lecture, Boston University April 22, 2019, https://www.bu.edu.
38 Hammonds, "Women in STEM in the Twenty-First Century."

CHAPTER 21

1 Banu Subramaniam, "And the Mirror Cracked," in *Feminist Science Studies: A New Generation*, ed. Maralee Mayberry, Banu Subramaniam, and Lisa H. Weasel (New York: Routledge, 2001), 55–57.
2 Subramaniam, "And the Mirror Cracked," 57.
3 Maralee Mayberry, Banu Subramaniam, and Lisa H. Weasel, "Adventures across Natures and Cultures: An Introduction," in *Feminist Science Studies*, 6.
4 Subramaniam, "And the Mirror Cracked," 59; Mayberry et al., "Adventures across Natures and Cultures," 6.
5 Subranamiam, "And the Mirror Cracked," 59–61.
6 Mayberry et al., "Adventures across Natures and Cultures," 10.
7 Mayberry et al., "Adventures across Natures and Cultures," 10–11.
8 Evelynn Hammonds and Banu Subramaniam, "A Conversation on Feminist Science Studies," *Signs* 28 (Spring 2003): 926.
9 Hammonds and Subramaniam, "A Conversation on Feminist Science Studies," 926.
10 "If I remember correctly, *Signs* invited me to interview Evelynn, and given Evelynn's generosity of spirit, it became a conversation rather than an interview" (Banu Subramaniam, personal communication, June 2, 2023).
11 Hammonds and Subramaniam, "A Conversation on Feminist Science Studies," 927–28.
12 Hammonds and Subramaniam, "A Conversation on Feminist Science Studies," 929.
13 Hammonds and Subramaniam, "A Conversation on Feminist Science Studies," 930.
14 Hammonds and Subramaniam, "A Conversation on Feminist Science Studies." 931–32.
15 Christa Kuljian, interview with Evelynn Hammonds, August 18, 2020.
16 Hammonds and Subramaniam, "A Conversation on Feminist Science Studies," 939.
17 Hammonds and Subramaniam, "A Conversation on Feminist Science Studies," 939–40.
18 Hammonds and Subramaniam, "A Conversation on Feminist Science Studies," 934–35.
19 Hammonds and Subramaniam, "A Conversation on Feminist Science Studies," 942.
20 Hammonds and Subramaniam, "A Conversation on Feminist Science Studies," 943.

EPILOGUE

1 Nidhi Subbaraman, "Inspired Choice," *Nature*, January 21, 2021, https://www.nature.com.
2 See Alondra Nelson's website, http://www.alondranelson.com.
3 Hannah Devlin, "Men Are Much More Likely to Die of Coronavirus—But Why?," *The Guardian*, April 16, 2020; Takehiro Takahashi, Mallory K. Ellingson, Patrick Wong, Benjamin Israelow, Carolina Lucas, Jon Klein, et al., "Sex Differences in Immune Responses That Underlie COVID-19 Disease Outcomes," *Nature*, August 26, 2020, 315–20; Roni Caryn Rabin, "As Women Prove Resilient to Virus, Trials Test Hormones on Men," *New York Times*, April 27, 2020.
4 Ann Caroline Danielsen, Katharine M. N. Lee, Marion Boulicault, Tamara Rushovich,

Annika Gompers, Amelia Tarrant, et al., "Sex Disparities in COVID-19 Outcomes in the United States: Quantifying and Contextualizing Variation," *Social Science and Medicine* 294 (February 2022): 114716.

5 Evelynn Hammonds, "A Moment or a Movement: The Pandemic, Political Upheaval, and Racial Reckoning, *Signs* 47 (Autumn 2021): 11; Isaac Chotiner, "How Racism Is Shaping the Coronavirus Pandemic," *New Yorker*, May 7, 2020, https://www.newyorker.com.

6 Rachel E. Gross, "Opinion: Feminist Science Is Not an Oxymoron," *Undark*, September 15, 2022, https://undark.org.

7 See the Science for the People website, https://scienceforthepeople.org/.

8 "Which Way for Science," *Science for the People*, April 18, 2017, https://scienceforthe people.org.

9 Sexual harassment in the academy and in science is a serious issue for women. Evelynn Hammonds had direct dealings with sexual harassment policy on the Harvard campus in 2012 when she was serving as dean. See Mercer R. Cook and Rebecca D. Robbins, "Sexual Assault Policy Unlikely to Change Soon," *Harvard Crimson*, November 16, 2012. Hammonds was a member of the Committee on Women in Science, Engineering, and Medicine, which contributed to Paula A. Johnson, Sheila E. Widnall, and Frazier F. Benya, eds., *Sexual Harassment of Women: Climate, Culture, and Consequences in Academic Sciences, Engineering, and Medicine* (Washington, DC: National Academies Press, 2018). In February 2022, she was one of thirty-eight Harvard faculty members who signed a letter in support of anthropology professor John Comaroff, who had been sanctioned by Dean Claudine Gay and found to have violated Harvard's sexual and professional conduct policies. Henry Louis Gates, Jamaica Kincaid, Jill Lapore, and Maya Jasanoff were also among the signatories. But after three students filed a federal law suit against Comaroff and Harvard, most of the signatories, including Hammonds, retracted their signatures. During an online book talk with Anita Hill, for which Hammonds was the moderator, she apologized. "This has been a difficult soul-searching moment," said Hammonds. "After receiving additional information, I retracted my name from the statement. And I want to state that I absolutely stand with the students in this case. And I want to state that I completely support them and I applaud their enormous courage in bringing their charges forward and making these charges known to us all" (Darley A. C. Boit, Caroline E. Curran, and Sara Dahiya, "At Anita Hill Book Talk, Former Harvard College Dean Says She Regrets Signing Comaroff Letter," *Harvard Crimson*, February 11, 2022).

10 Alondra Nelson, "Weapons for When Bigotry Claims Science as Its Ally," *Nature*, September 7, 2020, 182–83

11 Chanda Prescod-Weinstein, "Making Black Women Scientists under White Empiricism: The Racialization of Epistemology in Physics," *Signs* 45, no. 2 (2020): 423, 427.

12 Chanda Prescod-Weinstein, *The Disordered Cosmos: A Journey into Dark Matter, Spacetime, and Dreams Deferred* (New York: Bold Type Books, 2021).

13 Christa Kuljian, interview with Banu Subramaniam, October 20, 2022.

14 Sophie Wang, *Science under the Scope: The Objectivity of My Affection, Full Series*, n.d., in *Free Radicals*, https://freerads.org; Sophie Wang and Alexis Takahashi, "The Origin Story of *Free Radicals*," January 9, 2021, https://freerads.org.

15 Emily Martin, personal communication, January 20, 2021.

16 Founders of *Catalyst* included Banu Subramaniam, Lisa Cartwright, David Serlin, Deboleena Roy, Elizabeth Wilson, Michelle Murphy, Rachel Leela, and Mara Mills. Evelynn Hammonds sits on the advisory board.

17 "Foundations and Futures of Feminist Technosciences," online celebration of *Catalyst*'s five-year anniversary, April 22, 2021, https://catalystjournal.org.

18 "Foundations and Futures of Feminist Technosciences."

19 "Foundations and Futures of Feminist Technosciences."

APPENDIX I

1 Ruth Hubbard and Elijah Wald, *Exploding the Gene Myth* (Boston: Beacon, 1993), 6.

APPENDIX 2

1 "Equality for Women in Science," *Science for the People* 2 (August 1970): 10–11.

2 The full text also appears in Rita Arditti, Pat Brennan, and Steve Cavrak, eds., *Science and Liberation* (Boston: South End Press, 1980), 283–86.

SELECTED BIBLIOGRAPHY

FROM THE 1970S

Arditti, Rita. "Women's Biology in a Man's World: Some Issues and Questions." *Science for the People* 5, no. 4 (1973): 39–42.

_____. "Women as Objects: Science and Sexual Politics." *Science for the People* 6, no. 5 (1974): 8–11, 28–31

Boston Women's Health Book Collective. *Our Bodies, Ourselves.* New York: Simon and Schuster, 1976.

Burnham, Dorothy C. "Biology and Gender." In *Genes and Gender*, edited by Ethel Tobach and Betty Rosoff, 51–59. New York: Gordian, 1978.

Hopkins, Nancy. "The High Price of Success in Science." *Radcliffe Quarterly* 10 (June 1976): 16–18.

Hubbard, Ruth. "Have Only Men Evolved." In *Women Look at Biology Looking at Women: A Collection of Feminist Critiques*, edited by Ruth Hubbard, Mary Sue Henifin. and Barbara Fried, 7–36. Cambridge, MA: Schenkman, 1979.

Hubbard, Ruth, Mary Sue Henifin and Barbara Fried, eds. *Women Look at Biology Looking at Women: A Collection of Feminist Critiques* Cambridge, MA: Schenkman, 1979.

Hubbard, Ruth, and Marian Lowe, eds. *Genes and Gender II: Pitfalls in Research on Sex and Gender.* New York: Gordian, 1979.

Keller, Evelyn Fox. "The Anomaly of a Woman in Physics." In *Working It Out*, edited by Sara Ruddick. New York: Pantheon, 1977.

Lowe, Marian. "Sociobiology and Sex Differences." *Signs* 4 (Autumn 1978): 118–25.

Malcom, Shirley Mahaley, Paula Quick Hall, and Janet Welsh Brown. "The Double Bind: The Price of Being a Minority Woman in Science." Report no. 76-R-3. Washington, DC: American Association for the Advancement of Science, Office of Opportunities in Science, 1976.

Salzman, Freda. "Are Sex Roles Biologically Determined?" *Science for the People* 9, no. 4 (1977): 27–32, 43.

———. "Sociobiology: The Controversy Continues." *Science for the People* 11, no. 2 (1979): 20–27.

Sayre, Anne. *Rosalind Franklin and DNA*. New York: Norton, 1975.

Smith, Barbara, Beverly Smith, and Demeta Frazier. "The Combahee River Collective Statement." In *Capitalist Patriarchy and the Case for Socialist Feminism*, edited by Zillah R. Eisenstein. New York: Monthly Review Press, 1978.

FROM THE 1980S

Arditti, Rita, Pat Brennan, and Steve Cavrak, eds. *Science and Liberation*. Boston: South End Press, 1980.

Bleier, Ruth. *Science and Gender: A Critique of Biology and Its Theories on Women*. Oxford: Pergamon, 1984.

———, ed. *Feminist Approaches to Science*. Oxford: Pergamon, 1986.

Fausto-Sterling, Anne. "The Myth of Neutrality: Race, Sex, and Class in Science." *Radical Teacher* 19 (1981): 21–25.

———. *Myths of Gender: Biological Theories about Women and Men*. New York: Basic Books, 1985.

Fausto-Sterling Anne, and Lydia L. English. "Women and Minorities in Science: An Interdisciplinary Course." *Radical Teacher* 30 (January 1986): 16–20.

Gilbert, Scott. *Developmental Biology*. Sunderland, MA, and Oxford: Sinauer Associates and Oxford University Press, 1985.

Hammonds, Evelynn, and Aimee Sands. "Never Meant to Survive: A Black Woman's Journey: An Interview with Evelynn Hammonds." *Radical Teacher* 30(January 1986): 8–15.

Haraway, Donna. *Primate Visions: Gender, Race, and Nature in the World of Modern Science*. New York: Routledge, 1989.

hooks, bell. *Ain't I a Woman?* Boston: South End Press, 1981.

Hubbard, Ruth. "Facts and Feminism." *Science for the People* 18, no. 2 (1986): 17–20, 26.

———. "Reflections on My Life as a Scientist." *Radical Teacher* 30 (January 1986): 3–7.

Hubbard, Ruth, Mary Sue Henifin, and Barbara Fried, eds. *Biological Woman— The Convenient Myth: A Collection of Feminist Essays and a Comprehensive Bibliography*. Cambridge, MA: Schenkman, 1982.

Hull, Gloria T., Patricia Bell Scott, and Barbara Smith, eds. *All the Women Are White, All the Blacks Are Men, But Some of Us Are Brave*. New York: Feminist Press, 1982.

Keller, Evelyn Fox. *A Feeling for the Organism: The Life and Work of Barbara McClintock*. New York: Freeman, 1983.

_____. "Feminism and Science." *Signs* 7 (Spring 1982): 589–602.

_____. "The Gender/Science System: Or Is Sex to Gender as Nature Is to Science?" *Hypatia* 2 (September 1987): 37–49.

_____. *Reflections on Gender and Science.* New Haven, CT: Yale University Press, 1985.

Lorde, Audre. *The Cancer Journals.* San Francisco: Aunt Lute Books, 1980.

_____. *Sister Outsider: Essays and Speeches.* Berkeley, CA: Crossing, 1984.

Manning, Kenneth. *Black Apollo of Science: The Life of Ernest Everett Just.* Oxford: Oxford University Press, 1985.

Martin, Emily. *The Woman in the Body: A Cultural Analysis of Reproduction.* Boston: Beacon, 1987.

Merchant, Caroline. *The Death of Nature: Women, Ecology, and the Scientific Revolution.* San Francisco: Harper and Row, 1980.

Moraga, Cherrie, and Gloria Anzaldúa, eds. *This Bridge Called My Back: Writings by Radical Women of Color.* Watertown, MA: Persephone, 1981.

Pearson, Willie. *Black Scientists, White Society, and Colorless Science: A Study of Universalism in Science,* Millwood, NY: Associated Faculty Press, 1985.

Rossiter, Margaret W. *Women Scientists in America.* Volume 1, *Struggles and Strategies to 1940.* Baltimore: Johns Hopkins University Press, 1982.

Schiebinger, Londa. *The Mind Has No Sex? Women in the Origins of Modern Science.* Cambridge, MA: Harvard University Press, 1989.

Smith, Beverly. "Black Women's Health: Notes for a Course." In *Biological Woman—The Convenient Myth: A Collection of Feminist Essays and a Comprehensive Bibliography,* edited by Ruth Hubbard, Mary Sue Henifin, and Barbara Fried, 103–14. Cambridge, MA: Schenkman, 1982.

Traweek, Sharon. *Beamtimes and Lifetimes: The World of High Energy Physics.* Cambridge, MA: Harvard University Press, 1988.

White, E. Frances, and Ann Woodhull-McNeal, "Challenging the Scientific Myths of Gender and Race: A Review of *Myths of Gender.*" *Radical America* 20, no. 4 (1986): 25–32.

FROM THE 1990S

Birke, Lynda, and Ruth Hubbard, eds. *Reinventing Biology: Respect for Life and the Creation of Knowledge.* Bloomington: Indiana University Press, 1995.

Collins, Patricia Hill. *Black Feminist Thought.* New York: Routledge, 1990.

Fausto-Sterling, Anne. "Building Two-Way Streets: The Case of Feminism and Science." *National Women's Studies Association Journal* 4, no. 3 (1992): 336–49.

Ferree, Myra Marx, Judith Lorber, and Beth B. Hess. *Revisioning Gender.* Walnut Creek, CA: Altamira, 1998.

Hammonds, Evelynn. "Black (W)holes and the Geometry of Black Female Sexuality." *differences* 6, nos. 2–3 (1994): 126–45.

_____. *Childhood's Deadly Scourge: The Campaign to Control Diphtheria in New York City*. Baltimore: Johns Hopkins University Press, 1999.

Harding, Sandra. "Rethinking Standpoint Epistemology: What Is 'Strong Objectivity?'" *Centennial Review* 36, no. 3 (1992): 437–70.

_____, ed. *The "Racial" Economy of Science: Toward a Democratic Future*. Bloomington: Indiana University Press, 1993.

Hubbard, Ruth. *The Politics of Women's Biology*. New Brunswick, NJ: Rutgers University Press, 1990.

Hubbard, Ruth, and Elijah Wald. *Exploding the Gene Myth*. Boston: Beacon, 1993.

Kass-Simon, G., and Patricia Farnes, eds. *Women of Science: Righting the Record*. Bloomington: Indiana University Press, 1993.

Keller, Evelyn Fox, and Marianne Hirsch, eds. *Conflicts in Feminism*. New York: Routledge, 1991.

Keller, Evelyn Fox, and Helen Longino, eds. *Feminism and Science*. Oxford: Oxford University Press, 1996.

Kohlstedt, Sally Gregory, and Helen Longino. "The Women, Gender, and Science Question: What Do Research on Women in Science and Research on Gender and Science Have to Do with Each Other?" *Osiris* 12 (1997): 3–15.

Longino, Helen, and Evelynn Hammonds. "Conflicts and Tensions in the Feminist Study of Gender and Science." In *Conflicts in Feminism*, edited by Evelyn Fox Keller and Marianne Hirsch, 164–83. New York: Routledge, 1990.

Martin, Emily. "The Egg and the Sperm: How Science Has Constructed a Romance Based on Stereotypical Male-Female Roles." *Signs* 16, no. 3 (1991): 485–501.

Noble, David. *A World without Women: The Christian Clerical Culture of Western Science*. New York: Knopf, 1993.

Oreskes, Naomi. "Objectivity or Heroism? On the Invisibility of Women in Science." *Osiris* 11 (1996): 87–113.

Roberts, Dorothy. *Killing the Black Body: Race, Reproduction, and the Meaning of Liberty*. New York: Vintage, 1997.

Rossiter, Margaret W. "The Matthew Matilda Effect in Science." *Social Studies of Science* 23, no. 2 (1993): 325–41.

_____. *Women Scientists in America*. Volume 2, *Before Affirmative Action, 1940–1972*. Baltimore: Johns Hopkins University Press, 1998.

Schiebinger, Londa. *Nature's Body: Gender in the Making of Modern Science*. New Brunswick, NJ: Rutgers University Press, 1993.

FROM THE 2000S

Beckwith, Jonathan. *Making Genes, Making Waves: A Social Activist in Science.* Cambridge, MA: Harvard University Press, 2002.

Collins, Patricia Hill. "Moving beyond Gender: Intersectionality and Scientific Knowledge." In *Revisioning Gender,* edited by Myra Marx Ferree, Judith Lorber, and Beth B. Hess. Walnut Creek, CA: Altamira, 2001.

Fausto-Sterling, Anne. *Sexing the Body: Gender Politics and the Construction of Sexuality.* New York: Basic Books, 2000.

Hammonds, Evelynn, and Rebecca Herzig, eds. *The Nature of Difference.* Cambridge, MA: Massachusetts Institute of Technology Press, 2009.

Hammonds, Evelynn, and Banu Subramaniam. "A Conversation on Feminist Science Studies." *Signs* 28, no. 3 (2003): 923–44.

Hubbard, Ruth. "Power, Gender: How DNA Became the Book of Life." *Signs* 28, no. 3 (2003): 791–99.

Keller, Evelyn Fox. *The Century of the Gene.* Cambridge, MA: Harvard University Press, 2000.

Lederman, Muriel, and Ingrid Bartsch, eds. *The Gender and Science Reader.* New York: Routledge, 2000.

Mayberry, Maralee, Banu Subramaniam, and Lisa H. Weasel, eds. *Feminist Science Studies: A New Generation.* New York: Routledge, 2001.

Reardon, Jenny. *Race to the Finish: Identity and Governancy in an Age of Genomics.* Princeton, NJ: Princeton University Press, 2005.

Schiebinger, Londa. *Has Feminism Changed Science?* Cambridge, MA: Harvard University Press, 2001.

Segerstrale, Ullica. *Defenders of the Truth: The Battle for Science in the Sociobiology Debate and Beyond.* Oxford: Oxford University Press, 2000.

Subramaniam, Banu. "And the Mirror Cracked." In *Feminist Science Studies: A New Generation,* edited by Maralee Mayberry, Banu Subramaniam, and Lisa H. Weasel. New York: Routledge, 2001.

_____. "Moored Metamorphoses: A Retrospective Essay on Feminist Science Studies." *Signs* 34, no. 4 (2009): 951–80.

_____. "Snow Brown and the Seven Detergents: A Metanarrative on Science and the Scientific Method." In *Women, Science, and Technology: A Reader in Feminist Science Studies,* edited by Mary Wyer, Mary Barbercheck, Donna Giesman, Hatice Orun Ozturk, and Marta Wayne. New York: Routledge, 2001.

Ulrich, Laurel Thatcher. *Yards and Gates: Gender in Harvard and Radcliffe History.* New York: Palgrave Macmillan, 2004.

Williams, Clarence G. *Technology and the Dream: Reflections on the Black Expe-*

rience at MIT, 1941–1999. Cambridge, MA: Massachusetts Institute of Technology Press, 2001.

Wyer, Mary, ed. *Women, Science, and Technology.* New York: Routledge, 2001.

FROM THE 2010S TO THE PRESENT

Bailey, Moya. *Misogynoir Transformed: Black Women's Digital Resistance.* New York: New York University Press, 2021.

Benjamin, Ruha. *People's Science: Bodies and Rights on the Stem Cell Frontier.* Stanford, CA: Stanford University Press, 2013.

_____. *Race after Technology: Abolitionist Tools for the New Jim Code.* Cambridge, UK: Polity, 2019.

_____. *Viral Justice: How We Grow the World We Want.* Princeton, NJ: Princeton University Press, 2022.

Braun, Lundy. *Breathing Race into the Machine.* Minneapolis: University of Minnesota Press, 2014.

Díaz, Sara P. "'A Racial Trust': Individualist, Eugenicist, and Capitalist Respectability in the Life of Roger Arliner Young." *Souls* 18, nos. 2–4 (2016): 235–62.

Hogarth, Rana. *Medicalizing Blackness: Making Racial Difference in the Atlantic World, 1780–1840.* Chapel Hill: University of North Carolina Press, 2017.

Keller, Evelyn Fox. *The Mirage of a Space between Nature and Nurture.* Durham, NC: Duke University Press, 2010.

_____. *Making Sense of My Life in Science: A Memoir.* Amherst, MA: Modern Memoirs, 2023.

Kuljian, Christa. *Darwin's Lunch: Science, Race, and the Search for Human Origins.* Johannesburg: Jacana, 2016.

Nelson, Alondra. *Body and Soul.* Minneapolis: University of Minnesota Press, 2011.

_____. *The Social Life of DNA: Race, Reconciliation, and Reparations after the Genome.* Boston: Beacon, 2016.

Prescod-Weinstein, Chanda. *The Disordered Cosmos: A Journey into Dark Matter, Spacetime, and Dreams Deferred.* New York: Bold Type Books, 2021.

_____. "Making Black Women Scientists under White Empiricism: The Racialization of Epistemology in Physics." *Signs* 45, no. 2 (2020): 421–47.

Richardson, Sarah. "Feminist Philosophy of Science: History, Contributions, and Challenges." *Synthese* 177, no. 3 (2010): 337–62.

_____. *The Maternal Imprint: The Contested Science of Maternal Fetal Effects.* Chicago: University of Chicago Press, 2021.

_____. *Sex Itself: The Search for Male and Female in the Human Genome.* Chicago: University of Chicago Press, 2013.

Roberts, Dorothy. *Fatal Invention: How Science, Politics, and Big Business Recreate Race in the Twenty-first Century*. New York: New Press, 2011.

Rossiter, Margaret W. *Women Scientists in America*. Volume 3, *Forging a New World Since 1972*. Baltimore: Johns Hopkins University Press, 2012.

Roy, Deboleena. *Molecular Feminisms: Biology, Becomings, and Life in the Lab*, Seattle: University of Washington Press, 2018.

Saini, Angela. *Inferior: How Science Got Women Wrong and the New Research That's Rewriting the Story*. Boston: Beacon, 2017.

Shetterley, Margot Lee. *Hidden Figures: The American Dream and the Untold Story of the Black Women Who Helped Win the Space Race*. New York: Morrow, 2016.

Subramaniam, Banu. *Ghost Stories for Darwin: The Science of Variation and the Politics of Diversity*. Urbana: University of Illinois Press, 2014.

Tallbear, Kim. *Native American DNA: Tribal Belonging and the False Promise of Genetic Science*. Minneapolis: University of Minnesota Press, 2013.

Taylor, Keeyanga-Yahmatta, ed. *How We Get Free: Black Feminism and the Combahee River Collective*. Chicago: Haymarket, 2012.

Wang, Sophie. *Science under the Scope*, no date. In *Free Radicals*. https://freerads.org/science-scope-full/

Zernike, Kate. *The Exceptions: Nancy Hopkins, MIT, and the Fight for Women in Science*. New York: Scribner, 2023.

INDEX

Hubbard, Ruth (*continued*)
Fausto-Sterling and, 136, 141–42; feminist critique of science and, 185; at Genes and Gender symposium, 95–97; Guggenheim fellowship, 20; and Hammonds, 132, 140; Harvard full professorship, 57; Harvard lecturer status, 54; Harvard Medical School consideration, 38; Harvard protest, 53–54; Harvard research associate, 28; Harvard University allegiance, 79; "Have Only Men Evolved?" 59, 87, 92, 131, 160; Human Genome Project and, 174–75; influence on women scientists, 155, 193, 206, 208, 216; and Keller, 115, 119–20; Keller and, 149–50; life of, 221; male biases and, 121; marriages and divorce of, 19, 28; Nazis and World War II, 18; non-scientists and, 115; Paul Karrer Medal with Wald, 28, 237n50; "The Politics of Health Care" course, 54; Radcliffe College experience, 17–19, 40; research in Wald's lab, 20–22, 30; retirement of, 151, 237n2; Rosalind Franklin biography review, 125; and Rossiter, 74, 137; Science for the People and, 60; scientists views of, 216; "The Second Sex—Thirty Years Later" seminar, 110, 114; on sociobiology, 245n8; on sociobiology and sexism, 76–78, 80–81, 83, 93; Sociobiology Study Group and, 94–95; and Sokal hoax, 185; students thoughts on discrimination, 207; and Summers, 206; tenured biology professor, 1, 233n1; on Wald's Nobel Prize, 29–30, 138, 183, 237n53; Washington conference on sociobiology (1978), 98; and Wellesley seminar, 129; *Women Look at Biology Looking at Women*, 86–87; women's movement and science, 43; writings used in seminars, 148, 166. *See also* Wald, Deborah; Wald, Elijah; Wald, George
Hull, Akasha Gloria, 106
Hull, Gloria, 130
Human Genome Project, 174–75, 188, 191, 207
Hypatia, 166

Im/partial Science (Spanier), 175
"The Importance of Feminist Critique for Contemporary Cell Biology," 166–67

Inferior (Saini), 216
"Innate Confusions" Radcliffe talk, 211
The Insect Societies (E. O. Wilson), 75–76
Institute for Advanced Studies, Princeton, 179–80
intersectionality, 7, 97, 131, 180, 182, 209, 211, 214, 217, 219

Jackson, Shirley, 64, 67–68, 224, 243n5, 243n22
Jensen, Arthur, 50, 54, 56, 77, 81, 106, 241n15
Johns Hopkins University, 164
Johnson, Katherine, 5
Johnson, Lyndon Baines, 32
Journal of Bacteriology, 26
Journal of Molecular Biology, 26
Journal of Virology, 189
Judaeo-Christian belief system, 88
Just, Ernest Everett, 138–39, 147, 167, 205

Katzir-Katchalsky, Aharon, 44
Keller, Evelyn Fox, 223, Fig. 10; adjunct professor at MIT, 149; "The Anomaly of a Woman in Physics," 72–73, 116; bias against women in physics, 22–23, 44, 236n23; at Brandeis University, 22–23; challenges facing women scientists, 173; family background, 6; and Fausto-Sterling, 136, 149; *A Feeling for the Organism* (Keller), 117–19; "Feminism and Science," 121; feminist critique and, 154; Francis Bacon and, 115; "Gender and Science" course, 114, 144; gender and science issues, 113–14, 121; Harvard experience, 22, 24; and Hammonds 70, 73, 144, 148–9; and Hopkins, 193; and Hubbard, 115, 149–50; Human Genome Project and, 174; influence on women scientists, 155, 182, 206, 208, 210, 216; "Innate Confusions" Radcliffe talk, 206; and Latour, 187; life of, 222; MacArthur fellowship, 186; and McClintock, 116–20; MIT committee and, 194; at MIT science, technology and society program, 113, 120, 186; at Northeastern University teaching mathematics, 120; oral history of, 27; race and racism, 217; retirement of, 188;